Small and Medium Sized Companies in Europe

Springer-Verlag Berlin Heidelberg GmbH

David M. W. N. Hitchens, Mary Trainor,
Jens Clausen, Samarthia Thankappan,
Bruna de Marchi

Small and Medium Sized Companies in Europe

Environmental Performance, Competitiveness and Management: International EU Case Studies

In collaboration with: Michael Keil (IÖW), Wilfried Konrad (IÖW) and Daniele Ungaro (ISIG)

Queen's University Belfast

Institut für ökologische Wirtschaftsforschung gGmbH
iöw

Springer

ISIG
ISTITUTO DI SOCIOLOGIA INTERNAZIONALE DI GORIZIA

D.M.W.N. Hitchens
M. Trainor
S. Thankappan
School of Management and Economics
Queen`s University,
Belfast BT71NN, United Kingdom

Jens Clausen
Institut für Ökologische
Wirtschaftsforschung (IOW), gGmbH
Hausmannstrasse 9-10
30159 Hannover, Germany

Bruna de Marchi
Head Mass Emergencies Programme
Institute of International Sociology of Gorizia (I.S.I.G.)
Via Mazzini 13, 34170 Gorizia, Italy

ISBN 978-3-642-07275-8 ISBN 978-3-662-09920-9 (eBook)
DOI 10.1007/978-3-662-09920-9

Library of Congress Cataloging-in-Publication Data Applied For

A catalog record for this book is available from the Library of Congress.
Bibliographic information published by Die Deutsche Bibliothek
Die Deutsche Bibliothek lists this publication in die Deutsche Nationalbibliographie; detailed bibliographic data is available in the Internet at <http://dnb.ddb.de>.

http://www.springer.de

Originally published by Springer-Verlag Berlin Heidelberg New York in 2003.

Softcover reprint of the hardcover 1st edition 2003

Cover Design: *Kirchner, Springer-Verlag* Heidelberg
Typesetting: Camera-ready by the athors

Printed on acid free paper 30/2132 AO – 5 4 3 2 1 0

Authors

David Hitchens is Professor of Applied Economics at Queen's University Belfast. His principal fields of interest are in environmental regulation, industrial productivity, competitiveness and policy.

Mary Trainor is a Senior Research Officer in the School of Management and Economics at Queen's University Belfast. She has published on the economic performance of SMEs and regional development. (*E-mail: m.trainor@qub.ac.uk*)

Samarthia Thankappan is a researcher at Queen's University Belfast. Her expertise is in the field of social and economic aspects of sustainability, and technological policy issues.

Jens Clausen is a diploma'd engineer, and is associated with the Institute for Ecological Economy Research (IÖW), Germany. He is actively involved in the research on environmental management.

Bruna De Marchi is a sociologist and the head of the Mass Emergencies Programme at the Institute of International Sociology in Gorizia (ISIG), Italy. She also teaches on Environmental Risk in the University of Treiste.

Michael Kiel is a researcher at the Institute for Ecological Economy Research (IÖW), Germany.

Wilfried Konrad is a researcher at the Institute for Ecological Economy Research (IÖW), Germany. His key research areas are system innovations, eco-efficient services and product-related environmental information systems.

Daniele Ungaro is an Associate Professor in Political Sociology at University of Teramo.

Acknowledgements

The authors would like to thank the European Commission for funding this study under the Human Dimensions of Environmental Change work programme of DG Research. In particular we would like to thank Michael Cornaert and Per Backe-Hansen.

The Italian team also wishes to thank Michela De Nardo, Bruno Maltoni, Francesca Odella, Silvia Fracaros, Francesca Dalla Tana and Serafin Corral Quinatana. The German team wishes to acknowledge the help of Björn Zapfel, Tobias Ott and Thomas Rabe. The UK team acknowledges the helpful assistance of Mairead McIvor, Rachel Hilliard, Dr Esmond Birnie, and Bill Watts.

The international teams of researchers have also liaised with industry and environmental associations, universities and government departments, we are grateful for their assistance. The study would not have been possible without the willing help of almost 1500 firms which completed questionnaires and agreed to long interviews in order to fulfil the objectives of this study, and those regulators, industrialists, academics and trade association members and environmentalists who contributed to focus groups.

Belfast, April 2003

Contents

1 Study Background

1.1 Aims and Objectives

The major aim of this research is to identify and weigh the importance of factors that promote and constrain, the adoption of environmental initiatives by small and medium sized enterprises (SMEs). The objective is to inform how policy can overcome obstacles so as to promote the adoption of cleaner technology (including environmentally sensitive products (Oosterhuis et al. 1996)) by industry. *Central to the research is the testing of a set of hypotheses, which, inter alia, relate the adoption of cleaner technologies to competitiveness, management culture and the importance of the provision of information.* The manufacturing sectors considered are those dominated by SMEs where product *and* process environmental response by the firm is important. Of these sectors, furniture, textile finishing, and fruit and vegetable processing were chosen. More specifically the focus is on European SMEs i.e. those employing less than 250 employees.[1] Variations within the EU with respect to environmental regulation are exemplified by a study of firms (and plants) across four member states: Germany, North-east Italy, Republic of Ireland and the United Kingdom.

The adoption of clean technologies has been slow and uneven and action has mainly involved good housekeeping. Major changes involving large capital spending, or material substitution, process redesign or reformulation have been relatively rare (OECD 1985, 1995; Ashford 1993). The problem is particularly acute amongst SMEs (and the majority of enterprises are SMEs; e.g., two-thirds in Germany) because they are additionally handicapped by a lack of information and resources to invest in cleaner technologies (OECD 1990). While the dominant influence on a company's investment in environmental initiatives is the need to comply with regulations (Irwin and Hooper 1992; Whitaker 1993; Department of the Environment 1994; Green et al. 1994; CIA 1993), survey data indicates that many businesses are not fully aware of their legal environmental responsibilities while, some managers go beyond the "compliance only" approach and see benefits to be derived from an effective environmental programme (Winter and Ledgerwood 1994).

The widespread use of clean technologies depends on more than just technological factors and the return on investment and this study focuses on three interrelated hypotheses that recur in the literature. These are concerned with the impact of clean technology on firm competitiveness (and the influence of firm competi-

[1] In this study an upper limit of 500 employees has been taken for technical reasons.

tiveness on the take up of clean technology); the relevance of management's environmental awareness/culture; and the availability and quality of external information sources, including those arising from relationships and linkages to the firm (namely, trade associations, universities, R&D organisations, official government sources, business to business networks, environmental consultants, customers, suppliers etc.).

1.2 SMEs and Clean Technologies[2]

Small and medium sized enterprises are important to all European economies. They create jobs, are a source of innovation and competition, create a dynamic, healthy market economy and preserve a stable economic base. Despite considerable efforts by governments to inform SMEs of the potential economic benefits from positively managing their environmental performance and investing in clean technologies, investigations and surveys, indicate that many are still:

- unaware of relevant legislation,
- unconvinced of the potential cost savings and market opportunities,
- out of step with their customer's requirements,
- disassociated from their stakeholders' concerns.

Indeed, most SMEs see no reason to address the environmental aspects of their businesses. Any action taken is often a response to legislative and regulatory pressures rather than positively seeking new opportunities from environmental management. SMEs appear passive or at best reactive on environmental issues, displaying none of the innovative characteristics associated with the sector.

The message does not even get through on the need to comply with environmental regulations. This can partly be explained by the fact that SMEs are unreceptive or unable to interpret the relevance of the legislation to their business. Moreover, SMEs are expected to have expertise in a whole raft of complex regulatory areas such as tax law, VAT and employment law. These can be much more pressing than environmental legislation, in part because they are more fundamental to the operation of the business and because they are enforced more effectively.

Consequently, SMEs appear powerless to do anything about their environmental performance and do not fully recognise their contribution to environmental problems. Few seek the help available because the support services do not match their needs. Furthermore, smaller businesses lack the time and money to investigate their environmental performance or access the high cost consultancy support network. Despite the ability of large businesses to link product quality and cost savings to environmental performance, SMEs lack the internal skills to respond to new demands or at the pace required. Finally and most importantly, SMEs lack the

[2] See Appendix S3, Clean technology adoption by firms, M Vasilopoulos in Hitchens et al. (2001). Sects. 1.4 and 1.5 also follow this Appendix.

financial potency or funds for R&D to keep abreast with technological improvements.

In general small companies:

- have limited financial resources,
- believe that environmental issues represent low priority,
- are unfamiliar with the concept of clean technologies,
- wait for compulsory regulation instead of anticipating future requirements,
- lack know-how as regards environmental regulation and technological developments.

While this may describe the average position of the environmental performance of SMEs, there are nevertheless cases where the use of clean technologies and the environmental performance of firms are above the average of SMEs in each industry.

This study focuses on the characteristics and experience of those above average environmental performers, in comparison with the average performers in the industry, and more specifically it seeks to test the following hypotheses to explain an above average take up of environmental initiatives.

1.3 Main Hypotheses

(i) The relationship between investment in environmental initiatives and firm competitiveness is likely to involve positive feedback in both directions

While clean technologies are expected to have consequences for the competitiveness of firms, the fact that the firm is characterised by above average competitive performance may lead to the early adoption of clean technologies.

The competitive performance of firms may be defined from the input or output side (Jacobson and Andréosso-O'Callaghan 1996). Where competitiveness is defined from the input side (i.e. representative of the likely explanations of competitiveness) the measure is based on strengths in physical and human capital endowment, R&D spending etc. These are also factors that are known to influence the adoption of cleaner production technologies (Green et al. 1994; OECD 1985, 1987, 1995; ECMT/OECD 1994; Wallace 1995). Furthermore the above average competitive performance of the firm implies that it also has the management capability to respond to environmental pressures with best practice solutions. These input hypotheses are considered in Sect. 1.3.

Output side indicators of competitive performance: profitability, market share, productivity, patents, firm growth etc., not only measure (in principle) the consequences of the adoption of clean technologies (ENDS 1994; CBI 1994; OECD 1987, 1995; Porter 1990), but also provides the resources and opportunities for the adoption of cleaner production methods and products.

Where investment in environmental initiatives arises through environmental regulation, there is no convincing evidence that this affects competitiveness (Cropper and Oates 1992; Jaffe et al. 1995; Glass 1996; Ekins and Speck 1998). Porter and Van der Linde (1995) go one step further to advocate the existence of benefits and substantial first mover advantages from the early application of environmental technologies. The question that arises here is whether those firms that go beyond compliance only (with regulation) depress or improve their competitive performance.

(ii) The role of management and the culture of the business organisation is important to the take up of environmental initiatives

Some managers are more likely than others to "internalise the externality" of environmental effects i.e. they will seek to avoid negative environmental outcomes even when these are purely external. Such management behaviour could result from an ethical commitment to valuing the environment *per se* (Etzioni 1988) and/or a market structure, which relax the constraint on firms to maximise profits and therefore allow the pursuit of a wider range of management goals (Williamson 1963). Alternatively the adoption of clean technology leads to improved profitability. Given that most firms are probably not profit maximisers and work within conditions of uncertainty a large part of company behaviour is probably governed by habit and rules (Simon 1962). Such rules of thumb and practices represent the culture of the company. The significance of the role of top management is important and an understanding of how they become convinced to choose cleaner technologies given the competitive position of the enterprise is crucial (O'Connor 1997).

Negative cultural influences on the take up of clean technology include a fear of change following the introduction of new technologies and the new organisational patterns which may arise (Christie et al. 1990). Further barriers are management inertia and a lack of internal communication (OECD 1995), and concerns among workers and unions with job losses (Smith 1977); although there is a lack of evidence of such job losses (OECD 1992a). There is conservatism and established practice in that most firms address specific pollution issues rather than strategically implementing clean technology investments to create a source of competitive advantage, i.e. for most firms their environmental policy is driven more by threat than opportunity (Newman et al. 1992).

It has been argued that of key importance in the achievement of sustainable business activity is a shift in management thinking, values and practice (O'Riordan 1985; Stead and Stead 1992; Roome 1994) so that the environment is placed high on the business agenda and environmental concerns are integrated into corporate culture (Winter and Ledgerwood 1994). In particular there is a need for top management support especially among SMEs (Hutchinson and Chaston 1994; Winter and Ledgerwood 1994; Christie et al. 1995; OECD 1985, 1987; Schmidheiny 1992) and, in addition, for that management to base its decisions on appropriate information (e.g. eco-audits, life cycle analyses) and accounting practices (Yakowitz 1991).

(iii) The influence of information and advisory sources

Imperfect information is inherent to the process of technological change and markets for information are notorious for being imperfect (Arrow 1962). In addition, the skill and know how of management and the labour force in SMEs is variable. It is therefore hypothesised that there is a need for knowledge and understanding (OECD 1987; Christie et al. 1995; BCC 1994; Staudenmaier 1985, Winter and Ledgerwood 1994) and independent advice (OECD 1987).

Moreover, economic theory suggests that the dissemination of information about new technology is likely to be a critical determinant of diffusion (Stoneman 1983). Information and advisory sources include: trade associations, universities, R&D organisations, official government sources, business to business networks, environmental consultants, help and advice from suppliers and customers. External company networks and information sources are critical for the firm to understand and respond to difficult problems such as environmental pressures (Pasquero 1991) and to guide company strategy, operations and R&D (Roome 1994). In fact, it has become increasingly recognised that business services, in particular, play a strategic role in enabling firms to achieve competitiveness by affecting adjustment to products, processes, skills, organisation and management in response to social, economic, institutional and legal pressures (Bailly 1987; Gillis 1987; Marshall 1988; Illeris 1989; Coffey and Polese 1989; Martinelli 1991; Perry 1991; Marshall and Wood 1992).

1.4 Other Hypotheses Considered

(iv) Modern machinery can embody good environmental performance

The age of the plant and machinery in each firm is likely to impact on environmental outcomes, costs of compliance and the number of clean technology initiatives. Broadly speaking, it could be anticipated that the more up-to-date equipment will embody the best environmental technology. The status quo is maintained when industrial facilities are old and capital turnover is low (Hartje and Lurie 1985). However, having such up-to-date relatively clean machinery could also mean that such a company did not require as many clean technology initiatives as a counterpart trying to come to terms with the implications of older and dirtier machinery.

(v) Skills will foster initiatives

Firms with an abundance of skills will undertake more initiatives to reduce environmental pollution. These firms are often more engaged in R&D and are likely to find ways to reduce pollution and introduce clean technologies (including through incremental innovation). The presence of skills is important, though it is also important to recognise that some clean technologies and innovation requirements, to deal with environmental effects, may need skills different from those necessary for the firm to compete in its usual line of business.

(vi) R&D will enable firms to take up technologies

The R&D activity of the firm is expected to play a critical role in the adoption and development of clean technologies and, because of its relationship with competitiveness, R&D is also seen as the means through which business develops environmentally benign products and processes (Kodama 1991; Christie et al. 1995; OECD 1995; Irwin and Hooper 1992; Roome 1994).

(vii) National regulation and grants foster environmental performance

Environmental initiatives are also hypothesised to be related to the stringency of environmental standards, levels of enforcement and environmental costs, and the relative importance of other drivers faced by the firm. Grants and subsidies are expected to raise the take up of clean technologies. The investigation undertaken here, by including firms at an EU-level (across a range of countries/regions), involves a variety of regulatory climates, costs and other drivers.

1.5 Important Constraints

Numerous impediments to the adoption of clean technologies have been identified and are discussed in OECD publications (1985, 1987, 1990, 1991, 1992a, 1992b, 1995) and other studies (Meller 1986; Commission of the European Communities 1983, 1984; PA Consulting Group 1991; ACOST 1992; Medhurst 1994).

1.5.1 Capital stock

Usually clean technologies (CTs) are more likely to be less costly in new plants than in old plants that require retrofitting to install this technology. In addition, an OECD study (1987) that was based on extensive interviews with industry shows that technical innovations are easier to achieve in industries in their investment renewal or modernisation phase. Moreover opportunities for investment in cleaner technologies are reduced where industry and firm growth rates are slow. Capital assets depreciation impedes changes in production processes. Existing equipment is rarely replaced if it has not been written off or if it has not reached the end of its useful life; which may substantially exceed a plant's depreciation period. This rigidity arises since most companies regard exiting equipment as a sunk cost and may continue to be beneficial even after the depreciation period is completed. Furthermore, CTs are sometimes more difficult to set up due to the specific characteristics required by each plant or company for their production process. This also hinders reselling ageing equipment once a more profitable replacement is on sale[3] (Hartje and Lurie 1985).

1.5.2 Clean Technologies vs. End-of-Pipe Technologies

Plant specific economics usually favour the introduction of End-of-Pipe (EOP) technologies for the following reasons.

- Cost advantage: EOP does not affect investments already in place.
- Risk averse: EOP techniques provide tested certainty regarding their efficiency. This risk averse attitude is very pronounced among SMEs.
- Emission specific regulatory rigidities may in effect require EOP technologies.
- Market pressure: capital markets along with banks or other lenders are often pushing forward conflicting demands (i.e. high rates of return while maintaining a good environmental profile). Therefore, it is often the case that companies tend to choose, after a shortsighted cost-benefit calculation, techniques that have the least cost and maximum benefit short term while overlooking cleaner and more profitable options that might have a longer payback period.
- Ratchet effect: the introduction of EOP limits the risk of regulations and standards being tightened up in the short to medium term.

1.5.3 Product quality

Changes to cleaner production techniques can affect quality and have repercussions downstream. Changing material inputs from more to less toxic substances, for example, necessitates careful consideration of how such changes will affect equipment performance and final product quality. In the same vein, product redesign may require equipment modifications within a process line.

1.5.4 Accounting obstacles

Access to satisfactory technical and management information is important. There may be difficulties in evaluating the profitability of environmental investments. Accurate costing involves the capabilities and data sources of a number of personnel including the materials manager, the production engineer, or the financial officer. Contributions from multiple, rather than, single departments will be necessary to assemble such data (White et al. 1992, 1995). Failure to accurately quantify benefits and costs could bias investment decisions against pollution prevention. Moreover, popular investment appraisal and accounting conventions, e.g. the sim-

[3] Innovation of new and more productive processes usually outpaces adoption capacity by industry.

ple pay back criterion, can argue against the adoption of cleaner technologies by obscuring their profitability (White and Becker 1991).

1.5.5 Capital rationing

Projects are desirable (profit-maximising) only if they have positive Net Present Values (NPVs). The cost of capital is the cost of not spending the $100 on something else. In fact, the most desirable projects are those that provide the highest NPV per dollar invested (Dixit and Pindyck 1994). Firms often set limits on the capital available to individual business units. This places a ceiling on funds available for new investment and forces managers to prioritise across projects, all of which may have positive NPVs (Antle and Eppen 1985).

1.5.6 Risk

There may be risk attached to the investment especially with novel technologies as distinct from known technologies. There may be a fear of a change to regulations and standards; supply side features of the pollution control industry are important, including speed of innovation (availability of clean technologies) and market constraints, tensions in favour of proven "conventional" processes (i.e. established technology versus novel solutions) and constraints arising from any inadequacy in the development of markets, e.g. for recovered or recycled products. Hence, if the project in question is riskier than the firm's business generally, the cost of capital will be higher, because riskier investments demand a higher return. There are difficulties in determining investments that are "risk-equivalent" which contributes to making the task subjective (Commoner 1994).

1.5.7 Other issues

Cleaner technologies are associated with certain contingent benefits, which are difficult to predict and quantify. Thus, incorporating the future monetary benefits of such an investment into a project profitability analysis is problematic. The monetary benefits depend on if, when, and how much liability cost is avoided (White and Becker 1991). Cleaner technology profits may materialise well beyond a certain time frame commonly applied in investment analyses. Analytical methods, which fail to capture this future stream of savings, contain inherent biases against cleaner technology investments. In the competition for limited capital resources, such investments are likely to be rendered non-competitive hence more traditional pollution control projects and projects that are primarily driven by non-environmental objectives are undertaken.

1.6 Technology Diffusion

Though not all the evidence on technology diffusion is conclusive, there is broad agreement on two points. First, new technologies are never adopted by all potential users at the same time. The widespread diffusion of new technologies can take anywhere from five to fifty years. Second, countless studies have confirmed that the diffusion of new technologies follows a predictable inter-temporal pattern i.e. technologies are adopted rather slowly at first, then more rapidly, and then slowly again as a technology specific "adoption ceiling" is reached. These stylised facts have prompted researchers to focus on two related questions: Why do some firms adopt a given innovation before others? Why do some innovations diffuse more quickly than others? (see Blackman and Bannister (1996) for an extensive analysis). Researchers have found that new technologies are adopted fastest by firms that are:

- large,
- have well-trained staff,
- incur high regulatory costs when using an existing technology,
- have infrastructure complementary to the new technology,
- are in fast-growing industries,
- invest more in R&D,
- pay relatively low prices for inputs used intensively by the new technology,
- have relatively old existing capital.

Despite considerable research, the evidence regarding the impact of market structure, the degree to which the market is competitive or controlled by a small number of firms and the timing of diffusion is inconclusive. Moreover, even if clean technologies have significantly lower production costs, diffusion will not be immediate. To this end, all firms will not necessarily rapidly adopt CTs simply because, in most cases, they reduce production costs. A broad range of firm-level, sector-level, and country-level characteristics determine whether or not and how quickly new technologies are adopted, and there are likely to be systematic differences between them in nearly all of these characteristics.

1.7 Research Outcomes

The research is aimed at policy makers at both the national and EU level as they attempt to devise policies that would enable the achievement of both competitiveness and environmental targets (and, to minimise the negative trade-offs between the two and perhaps even exploit any positive links between these goals). More particularly, while designing the research, the two anticipated benefits were:

- increased focus of both industrial and environmental policy instruments to improve their effectiveness and efficiency in achieving goals like "sustainability" and a higher "employment intensity" of growth and,

- better integration of industrial/regional policy at the national and EU levels with environmental policy (Commission of the European Communities 1993).

The research is relevant to Agenda 21 especially in the recognition that government alone cannot achieve the underlying principles. The commitment of relevant groups and people is necessary in order to bring about the objective of an integration of sustainability and growth. The fifth Community Action Programme of policy and action in relation to the environment and sustainable development also recognises that the road to sustainable development requires changes in behaviour in the business world and at the level of the ordinary citizen. The importance of information, education and training is stressed throughout the programme (De Marchi 1995).

1.8 Book Structure

The methodology and variables considered are described in Chap. 2. Chapter 3 provides a background to the three industrial sectors considered: furniture, textile finishing and fruit and vegetable processing. Chapters 4, 5 and 6 discuss the results of the research for each industrial sector. Chapters 7 and 8 consider the importance of external sources of advice and company culture respectively. Chapter 9 presents the conclusions and policy recommendations.

2 Method

This chapter discusses the research methodology and includes a description of the key variables, rationale for the chosen measures, a summary of the questionnaires and an outline of the analytical method. The focus of the research was fourfold:

1. to identify those factors which influence differences in the adoption of environmental initiatives among similar companies in three industries,
2. to identify costs and benefits arising from the adoption of environmental initiatives (including productivity, competitiveness and employment effects),
3. to investigate factors facilitating the adoption of clean technologies, including consultancy and other information services,
4. to identify from service suppliers the strengths and weaknesses encountered in the SME sector.

While these objectives are broad in nature, the study concentrated on three key determining variables of environmental performance and take up of clean technologies: namely, firm competitiveness, management culture and external advisory services.

2.1 Competitiveness

As we have seen, output indicators of competitive performance such as profitability, market share, productivity, patents and firm growth, not only measure (in principle) the consequences of environmental protection and the adoption of any clean technology or other environmental initiatives they also provide the resources and opportunities for the adoption of cleaner production methods and products. In addition, input definitions of competitiveness like physical and human capital endowment, R&D spending, age of machinery and levels of technology, are also factors which are known to influence the adoption of environmental initiatives. Hence the relationship between environmental initiatives and competitiveness is two-way. The fact that the firm is competitive may lead to the early adoption of environmental initiatives while at the same time environmental initiatives are expected to have consequences for the competitiveness of firms.

What then are appropriate measures of the competitive performance of firms? One such measure is profitability, but this depends on the type of market structure within which the firm is operating. This suggests that a single measure of competitiveness may be inappropriate thus; in this study a number of indicators are used.

These include output measures of performance such as profitability, productivity, labour growth; destination of sales particularly as regards exports; and input measures of performance corresponding to the input side of efficiency and competitiveness, for example, physical and human capital, R&D capability and age of machinery.[4]

The competitiveness effects of individual environmental initiatives on firm performance are considered in relation to the impacts on sales, costs, profits, and employment. With respect to the latter, this recognises the fear that the implementation of new (cleaner) technologies might displace existing workers and managers (OECD 1995).

2.2 Environmental Performance

There are various approaches to measuring environmental performance (James 1994; Welford 1994; Greenberg and Unger 1992) and in this study it was initially intended to base the measurement of firm environmental performance on both input and output indicators. However, it proved difficult for the respondents to quantify these output indicators thus the environmental performance measures used are based on environmental inputs and environmental management and procedures. In addition, the aim of constructing a set of environmental performance measures was:

- to evaluate the effort behind environmental management systems,
- to evaluate the number and quality of process oriented environmental protection activities and to take into account, to what degree the firm is aware of their success,
- to evaluate the range and quality of product oriented environmental protection activities and the related activities of marketing and market communication.

For each of these three main environmental activities, every firm interviewed was allocated points for the specific initiatives undertaken and these points were then subsequently aggregated to give a total score. In order to differentiate the better environmentally performing firms, these points were distributed across a wide spectrum of initiatives but within each area of activity there were a maximum number of points that could be scored. For example, if a firm separated more than six waste streams, it would be allocated two points for waste management since to award say, 10 points for 10 waste streams would have placed too much emphasis on the importance of waste management. Furthermore, it could generally be assumed that the most important waste streams will be separated first, the most important energy saving measures will be initiated first and so on.

The choice of environmental initiatives which were included in the research was based on an analysis of:

[4] The precise definitions of the variables are shown in the questionnaires, which are available from the authors on request.

- Reference documents on Best Available Techniques from the European IPPC Bureau for the respective sectors,
- The Pollution Prevention and Abatement Handbook of the World Bank,
- Environmental regulation focus in the participating countries,
- Environmental statements from firms, within the sectors, that applied EMAS.

Taking into consideration these sources, eight initiatives were chosen which were of high environmental importance as well as being reasonably commonplace in the three sectors. Nevertheless, some differences between the initiatives analysed in the three sectors remained. For example, solvent reduction was an important activity in furniture production but not in fruit and vegetable processing whereas water use and protection was an issue in fruit and vegetable processing but not in furniture manufacturing. Also, product related initiatives were generally not applicable in more than one sector. The initiatives included in the analysis and the detailed system of environmental performance measurement on the basis of the face-to-face firm interviews is shown in Tables 2.1–2.3. Points scored were divided between environmental process control, environmental management activities and ecological raw material and products (Table 2.4).

Table 2.1. Environmental management measurement

Initiative	Allocated Points			
	⅓	⅔	1	2
Environmental officer			Yes	
Environmental policy	Idea discussed	Draft version	Official version	
Environmental improvement programme[a]	Idea discussed	Draft version	Official version	
Environmental team	1 meeting per year	2 meetings per year	3+ meetings per year	
Evaluation of environmental performance of suppliers	Yes, for few suppliers	Yes, for many suppliers	Protocols for all suppliers	
Data collection for controlling environmental improvement[b]	Rarely	Quite often	Systematically	
EMAS or ISO14001 implemented or planned in the next 3 years			Yes	
EMS certified externally			Yes	
Environmental communication instruments			One	2 or more

[a]Excluding textile finishing.
[b]Excluding furniture.

Table 2.2. Environmental process control measurement

Initiative	Allocated Points 1	2
Solvent reduction initiatives[a]	One	2 or more
Environmental effect of solvent reduction[a]	Known	
Waste separation[b]	3–5 streams	6 or more streams
Environmental effect of waste separation	Known	
Energy saving initiatives	One	2 or more
Environmental effect of energy saving	Known	
Environmentally friendly packaging	One	2 or more
Environmental effect of env. friendly pack.	Known	
Water use and protection[c]	One	2 or more
Environmental effect of water protection[c]	Known	

[a]Excluding fruit and vegetable processing.
[b]Zero points for fewer than three separate waste streams.
[c]Excluding furniture.

Table 2.3. Ecological raw materials and products measurement

Initiative	Allocated Points 1	2
Furniture		
Ecological marketing strategy	Yes	
Natural/ecological materials	One	2 or more
Toxic use reduction in raw materials	One substance	2 or more
Eco-design scheme	Yes	
Design award	Yes	
Textile finishing		
Co-operation in environmental projects	One	2 or more
Toxic use reduction in raw materials	One substance	2 or more
Natural/ecological materials	One	
Eco-Tex standard 100 certification	Yes	
Fruit and vegetable processing		
Raw materials from organic farming	Some requirements of organic farming applied to supplies	100% certified organic raw material
Percentage of organic raw material processed	Up to 30%	Over 30%
Avoidance of artifical ingredients	One substance	2 or more
Organic products sold	Yes	
Regional sourcing	Yes	

Table 2.4. Maximum score possible

	Furniture	Textiles	Fruit & Vegetables
Environmental management	9	9	10
Environmental process control	12	15	12
Ecological raw material and products	7	6	8
Total	28	30	30

The measure of environmental performance that was constructed for the firms that responded to the postal survey was much simpler. In the postal survey, SMEs were asked to select from a list of 11 industry specific initiatives those that they undertook. Many of the initiatives, such as waste separation, energy efficiency, were common to all three sectors with some specialization e.g., regional sourcing in the fruit and vegetable sector, eco-design in the furniture sector. Using a simple count of the number of initiatives undertaken three performance groups were formed:

- *compliance only* for firms that undertook 0–2 initiatives,
- *compliance plus* for firms that undertook 3–5 initiatives,
- *excellence* for firms that undertook 6 or more initiatives.

The establishment of performance groups for the firms' interviewed face-to-face (long questionnaire) was more problematic. Early in the research it was recognised that under the schema developed by IÖW (i.e. firms with less than 13 points were defined as compliance only, those with 13–17 points as compliance plus and those with 18 or more points as excellent) most UK/Republic of Ireland firms would be categorised as achieving compliance only status. This made it difficult to compare compliance only with compliance plus firms in the UK/Republic of Ireland after matching SMEs on the basis of size, product/activity and where possible turnover. Hence a compromise was reached in that the IÖW method was adopted for the counting of initiatives but no absolute threshold for the number of initiatives between two matched pairs was set. It was clear from our initial interviews that in the UK/Republic of Ireland there were no SMEs that would meet the threshold of excellence.

2.3 Company Environmental Culture and Attitudes

In order to assess the influence of the internal culture of the firm as regards environmental values, a questionnaire[5] was designed to gauge management attitudes to environmental issues. This questionnaire was completed by the managers of the companies that participated in the face-to-face interviews hence it was possible to directly link management attitudes with firm environmental performance. A series of questions were asked in order to evaluate management's attitude to issues such

[5] Questionnaire available from authors on request.

as environmental concerns, government policy towards the environment and the effect of environmental policy on the individual firm. Attitudes were measured in a number of ways to crosscheck on the consistency of responses and there was no attempt made to ascertain the rationale for the attitude held.

2.3.1 Measuring firm culture

Firm culture was measured by considering the responses to the following eight questions and determining if the firm adopted a pro-environmental or anti-environmental attitude. The responses to the first four questions, measured on the Likert scale (1–4, Likert 1932), were used to determine the environmental culture of the firm.

Q1: To what extent do you personally agree with the following statements?
a. People worry too much about the fact that economic development damages the environment.
b. Economic development cannot be supported by the natural resources available.
c. Nowadays, full employment is more important than environment protection.
d. It is right to increase taxation for industries which pollute.

From statement a, the two degrees of agreement (fully agree and partially agree) denoted an *anti-environmental* attitude while the two degrees of disagreement (partially disagree and fully disagree) in the same statement indicated a *pro-environmental* culture. However, with the objective of checking the coherence of the responses, the scale in statement b was inverted whereby the levels of agreement signified a position of pro-environmentalism while the degrees of disagreement indicated anti-environmentalism. In statement c, the position of anti-environmentalism is demonstrated by the degrees of agreement while that of pro-environmentalism by the expression of degrees of disagreement. In statement d, the position of pro-environmentalism is very delicate because it means justifying an increase in taxes to achieve ecological objectives. In fact sometimes firms that had earlier indicated a pro-environmental attitude expressed a position of anti-environmentalism by recording their disagreement with this statement.

The responses to question Q2, which concentrates on energy conservation, and to question Q3, which focuses on waste reduction, were classified in a straightforward manner. Fully or partially agreeing to the statements represented a pro-environmental viewpoint whilst fully or partially disagreeing indicated an anti-environmental attitude.

Q2: There are various strategies by which government could encourage companies to use less energy. To what extent do you agree with each statement?
a. To invest public money in information campaigns, to encourage companies to use less energy.
b. To increase periodically taxes on energy consumption.
c. To increase taxes for companies which use an excessive amount of energy.
d. To reduce taxes for companies with low-energy production cycles.

Q3: There are various means by which the government could try to encourage companies in your sector to produce less waste so as to protect the environment. To what extent do you agree with each statement?
a. To invest public money in information campaigns to reduce waste.
b. To prohibit an excessive use of packaging for products, even if they become less attractive for the customer.
c. To increase taxes for companies which do not collect the products at the end of the life-cycle.
d. To reduce taxes for companies using recyclable materials.

The attitude of the firm to environmental legislation and its implementation was sought in question 4. Agreement with the first statement indicated a pro-environmental attitude. In the second statement, the opposite is true i.e. the declaration of agreement that environmental legislation does not adequately take into account the problems that can arise for firms from their adoption implies a position of (grudging) anti-environmentalism. In the final statement, disagreement signified that, other things being equal, implementation of environmental legislation does not have a serious impact on the preservation of jobs in the firm i.e. a pro-environmental attitude.

Q4: To what extent do you personally agree with each of these statements?
a. In a firm like yours it is easy enough to adapt the production cycle to meet the requirements of the environmental legislation.
b. In a firm like yours, environmental legislation does not consider the problems connected with the actual implementation of the rules.
c. In a firm like yours, it is impossible to achieve the two objectives of job preservation and respect of environmental legislation.

On the basis of the responses from these 15 statements, an overall measure of environmental attitude was calculated by assigning a score[6] to each statement and aggregating the scores into a single value. Hence, firms that scored 45 or above were classified as pro-environmental while those with scores of less than 45 were classed as anti-environmental.

The next question provided an insight into the attitude of the firms as regards environmental legislation and its implementation.

Q5: What in your opinion was the main impact of environmental legislation:
a. An incentive to improvement for the good of society.
b. A hindrance to market competitiveness.
c. Encourages increased efficiency of production.

Pro-environmentalist firms were those that considered legislation as an incentive, leading to improvements that benefited society as a whole while anti-environmentalist firms were those that consider legislation as an impediment to overall competitiveness. A third category, innovators, considered such legislation

[6] The scores ranged from 4 for a very positive response (i.e. pro-environmental) to a 1 for a very negative (i.e. anti-environmental) attitude.

as leading to higher efficiency through increased cost savings or market opportunities.

On the basis of question 6 it was possible to distinguish between the different degrees of environmentalism depending on the motivation for adopting environmental initiatives. Firms with a strong environmental ethos adopted environmental norms because they believed in the common interest (statement 2). Medium environmentalist firms regarded the market as the main driver for the adoption of the environmental initiatives while firms that adopted initiatives to comply with legislation were considered to have a low motivation for undertaking such initiatives.

Q 6: Many companies have contributed to a greater respect for the environment through energy savings, waste recycling, the non-use of pollutant substances and so on. Your firm adopted environmental initiatives because?
1. Of legislation.
2. It is in our interest to live in a clean environment.
3. Of new types of consumption now fashionable among our customers.

The organisational structure of a company and its relationship with environmental attitudes were derived from the following two questions. Question Q7 is related to the perception of the organisational model in the firm and question 8 gives an insight into the organizational culture of the company.

Q7: Which of the following principles do you believe to be more important for the success of your firm?
1. That employees conscientiously fulfil the tasks assigned to them.
2. That the best technical solution is always the target.
3. That individual responsibility is taken at each level of production.

Organisations with an authoritarian model (point 1) would generally have a low regard for environmental issues while firms with a participatory model (point 3) would be at the opposite end of the spectrum. A neutral or mid-way position was held by firms that have a technocratic model (point 2).

Q8: To what degree do you personally agree with each of these statements (scoring as on the Likert scale).
1. Company culture is formed by the values of the director/owner, management, shared by his/her assistants and employees.
2. Company culture depends upon the marketplace in which the company works.
3. Company culture is a combination of customer, employees and management values.

The recognition of management values as the principal cultural factor of the company (statement 1) may indicate an authoritarian organisational model. However, it is possible that the individual attitudes of management were an important factor in the adoption of environmental initiatives particularly in the context of the entrepreneurial firm. Company culture that depends on the market to drive the implementation of environmental initiatives usually corresponds with positive environmental attitudes since pressure from consumers and competition within the market, are strong innovative factors at an environmental level. The final statement conforms to a participatory organisational model and indicates the possibility of several sources of cultural environmental attitudes (family, school, and so on).

2.4 External Advisory Services

External advisory services are considered as those providing information to firms that will facilitate their improved environmental performance. This process of receiving and providing information was examined from both the firms' and the advice providers' perspective. In an attempt to assess the impact of external advice, firms were asked about the use of such services, the importance of support including grants, the importance of firm networks, supplier chain relationships, source of in-house and external expertise, involvement of technical expertise within the firm, support of top management, marketing and purchasing departments, availability of finance and environmental awareness.

Providers of advice were interviewed regarding their relationship with the firm, the breadth and depth of advice provided, the problems that firms may encounter in the take-up of advice, the effect advice has on the management culture of the firm.[7]

2.5 Research Study Method

There were three separate issues or focal points that the empirical work examined. Firstly, what is the incidence and breadth of use of clean technology by SME's and why are these technologies acquired? What is their economic impact? These issues are addressed in a *postal survey* using a brief questionnaire in each country.

The second matter was to examine in detail the environmental and economic performance of individual firms/plants, the management culture, processes of adoption and constraints especially with respect to information. These questions have been addressed in *face-to-face interviews* with manufacturing companies in the selected industries. Based on the hypotheses discussed in Chap. 1, the research compares the characteristics (Roome 1994) of those firms that:

- comply with regulation,
- go beyond compliance (compliance plus),
- achieve excellence.

The third focus was on sources of information/advice as coming from customers, suppliers, consultants and other advisory sources. This issue was examined in the face-to-face interviews with users i.e. the sample manufacturing firms. In addition *interviews* were conducted with suppliers of information i.e. industry associations, consultancies, suppliers, other companies and (sectoral, geographic and supply

[7] Questionnaires are available from the authors on request.

chain) networks, customers, universities, R&D organisations, government departments, journals and trade magazines etc.

2.5.1 Choice of country and industry

The countries selected for the study were the United Kingdom (UK), the Republic of Ireland (ROI), Germany and Italy. The justification for this set of sample countries is provided by the extent to which they vary with respect to both environmental regulation and productivity. There was broad coverage in all countries except Italy where the investigation concentrated in the North-East[8], particularly the three regions of Veneto, Friuli-Venezia Giulia and Trentino Alto-Adige (usually defined as Triveneto). The UK data (including Northern Ireland) and the ROI were aggregated since there was insufficient variation in economic performance and environmental requirements to justify treating them as separate jurisdictions.

The three industrial sectors that were selected for investigation were furniture, textile finishing, and fruit and vegetable processing. These three sectors have environmental characteristics that cut across the majority of areas where environmental protection is relevant. They were also industries where SMEs are important and industries which have a significant EU presence in each of the chosen countries. Simultaneously these countries represented variations in environmental regulatory stringency and enforcement, and economic performance of the sectors studied (the latter is considered in Chap. 3).

Germany is the leading furniture producer in Europe, followed by Italy and both countries are among the top 10 furniture exporters in the world. The largest textile industry in the European Union (EU) is in Italy. Food and drink is the largest industrial sector in the EU with production equal to approximately €600 bn. The importance of the EU fruit and vegetable processing and preserving sub-sector is small relative to other sub-sectors in the industry. Germany had the largest output share of the EU member states in 1997 accounting for 18.4% of total EU production, followed by the UK with a share of 17.1%, Italy had a 14.7% share and Ireland had the lowest output share accounting for only 0.55%.

2.5.2 Sample sizes

Table 2.5. Postal survey responses by country and sector

	UK/ROI	Germany	Italy	Total
Furniture	105	93	100	298
Textiles	78	73	100	251
Fruit & vegetable processing	87	108	100	295
Total	270	274	300	844

[8] The economy of North-East Italy is characterised by traditional manufacturing industry including textiles, clothing, footwear, leather and furniture.

The postal survey yielded a total of 844 responses, which represents a response rate of 14% in the UK/ROI and 15% in Germany. Italian interviews were conducted by telephone and achieved a substantially higher response rate of 50%.

Table 2.6. Face-to-face interviews with SMEs

	UK/ROI	Germany	Italy	Total
Furniture	32	33	33	98
Textiles	33	33	33	99
Fruit & vegetable processing	33	34	30	97
Total	98	100	96	294

Table 2.7. Interviews with advisory sources

	UK/ROI	Germany	Italy	Total
Furniture	22	36	24	82
Textile	18	44	32	94
Fruit & vegetable processing	22	21	52	95
General[a]	38	11	–	49
Total	100	112	108	320

[a] Advisors to more than one sector.

2.5.3 Focus groups

Focus group meetings were held in each country for each of the sectors and for the providers of advice. The purpose of these meetings was to address the hypotheses, and particularly those relating to management culture, competitiveness and the adequacy of the provision of advice. The constraints on adoption of environmental initiatives were also discussed. The participants in the industry focus groups included representatives from firms in the sectors in question, consultants, suppliers, customers, universities and government support agencies.

2.6 Questionnaires

The questionnaires used in the surveys have been referred to in the earlier sections of this chapter and the content of each is described briefly below.

2.6.1 Postal questionnaire

A short two-page questionnaire was developed for the postal survey. One page was devoted to the economic characteristics and performance of the firm, environmental strategy and environmental attitudes. The second page concentrated on the environmental initiatives adopted, their drivers, economic effects, obstacles and facilitators to adoption.

2.6.2 Face-to-face questionnaire to manufacturers

For the face-to-face interviews, with manufacturing firm managers, a longer semi-structured questionnaire was developed. This included questions on the economic characteristics and performance of the firm. Questions on environmental initiatives and performance made up the majority of the interview. Individual environmental initiatives undertaken were considered along with the required investment and running costs, environmental impact, effect on labour, sales and so forth. Constraints on the take up of initiatives and the importance of different sources of advice were also addressed.

2.6.3 Face-to-face questionnaire to advisory sources

For each sector the providers of advice were interviewed. Advisers were asked to describe the type of advice given and the typical relationship with an SME client firm. In addition, a typical advisory input and its impact on the client firm was examined, as well as how the firm collaborated, the drivers for use of the service and so on. Specialist questions were included for suppliers of environmental equipment, environmental consultants, public institutions and non-profit organisations. Also included were questions relating to the difficulties encountered in the take up of initiatives e.g. whether they were management or organisation focused.

2.6.4 Cultural attitudes questionnaire to manufacturers

The respondent to the manufacturer's face-to-face interview completed the culture questionnaire, which included questions on the attitude of management to matters such as environmental concerns, government policy towards the environment and the effect of environmental policy on the firm.

2.7 Analysis

2.7.1 Postal questionnaire data

Hypotheses were tested for each industry and for each country. Findings from the postal questionnaire were contrasted with those from the face-to-face manufacturing questionnaire. A number of hypotheses were also tested by pooling the UK/ROI and German data.[9]

2.7.2 Face-to-face manufacturers' data

Matched firm comparisons

A matched comparison was made between firms achieving different levels of environmental performance (compliance only, compliance plus, excellence). Face-to-face interviews took place with firms, matched primarily by size and product. Data were analysed for each industry and each country. There was variation in the quality of matching of compliance only with compliance plus and excellence firms between sample countries.

National comparisons

National comparisons between firms classified according to environmental performance: compliance only, compliance plus and excellence, aimed at estimating within country relationships between environmental initiatives and the set of hypothesised explanatory variables thereby holding constant regulation, environmental input prices (materials, labour etc.) and enforcement.

International comparisons

The findings were contrasted internationally. In addition, a number of analyses are based on the pooled UK/ROI and German data.

[9] It was not possible to combine the responses to the Italian short questionnaire with those for UK/ROI and Germany because of constraints on the Italian data.

3 Industrial Sectors Considered

The study focused on three industrial sectors, namely, furniture, textile finishing, and fruit and vegetable processing. This chapter considers the economic structure and performance of each of these sectors in the sample countries (Germany, Italy, UK, ROI) and the European Union (EU) as a whole.

3.1 Furniture Industry

In 1999, furniture production in the EU was worth approximately €74 bn and accounted for 1.9% of manufacturing output. Output had grown from just under €50 bn in 1989, an annual growth rate of 4.3% which was marginally above the total manufacturing average. This aggregate growth rate masks the sharp fall in output and value added that occurred in 1993 and the more moderate reductions during a number of subsequent years, as well as the high rates of annual growth in 1989 (10.4%) and 1990 (9.1%). Table 3.1 shows that the furniture sector in the sample countries, especially the UK and ROI, performed well during the 1990s.

Table 3.1. Furniture output (€ billion)

	1992	1995	1998
Germany	18.4	19.8	20.3
Italy	10.8	10.3	12.7
UK	6.3	6.1	9.9
ROI	0.21	0.28	0.36

Source: Panorama of European Industry, 2000.

The European furniture industry is dominated by small to medium sized companies with the ten largest European furniture companies providing less than 9% of total European furniture supply. In each of the sample countries (excluding Germany as comparable data is not available), very small firms (1–9 employees) make up the majority of enterprises, with 87% of enterprises employing fewer than 10 employees in Italy, 76% in the UK and 50% in ROI. In both the UK and ROI there was an absence of firms employing more than 500 persons (Table 3.2).

Table 3.2. Distribution of enterprises by size, 1997

	Germany	Italy	UK	ROI
1–9	Not available	31,038	5,723	154
10–49	Not available	4,322	1,315	131
20–49	605	1,412	527	63
50–99	340	331	226	15
100–249	305	116	174	6
25—499	89	23	59	0
500+	49	6	0	0
Total	1,388	35,836	7,529	306

Source: Panorama of European Industry, 2000.

In terms of exports, Italy is the leading exporter with 61% of its total furniture output being sold abroad. The comparable figures for Germany, UK and ROI are 21%, 15% and 28% respectively. Italy is also the only one of the four countries to operate a trading surplus (Table 3.3).

Table 3.3. Trade balance, furniture industry, 1999 (€ million)

	Exports	Imports	Trade balance
Germany	4,332	5,974	-1,642
Italy	7,710	825	6,885
UK	1,451	2,895	-1,445
ROI	104	249	-145

Source: Panorama of European Industry, 2000.

Comparing the national economic indicators with the EU as a whole (Table 3.4) shows that labour productivity in this sector is well above the EU average in Germany and higher than the EU average in both the UK and Italy. In ROI, labour productivity is substantially below this average. Despite the very small absolute size of the industry in Ireland, there was considerable output growth between 1987 and 1997 (annual average rate of 12.3%) indicating that this industry is developing and has achieved a rate of turnover per employee similar to that in the UK.

Table 3.4. Economic indicators, furniture industry, 1997

	Germany	Italy	UK	ROI	EU
Employment ('000)	168.0	92.9	137.5	5.6	742.0
Turnover (ECU bn)	20.4	12.6	10.3	0.4	76.0
Production value (ECU bn)	19.3	12.1	9.9	0.4	68.6
Value added (ECU bn)	6.9	3.2	4.0	0.1	23.7
Labour productivity ('000 ECU per employee)	41.1	34.5	37.4	26.4	32.0
Turnover per head (ECU '000)	121.4	135.6	74.9	71.4	102.4

Source: Panorama of EU Industry, 2000.

3.1.1 United Kingdom

For the UK as a whole, the manufacture of "other furniture" which comprises bedroom, living room and garden furniture and the finishing of furniture such as spraying, painting and upholstering was the most important sub-sector in terms of the number of enterprises, employment and value (Table 3.5). Chairs and seats also played an important role, with about 28% of all employment in the furniture industry, and accounting for about 28% of the total production value, a proportion similar to the other furniture category. Office and shop furniture was the next most important sub-sector, accounting for 17% of total enterprises in the sector, 21% of employment and 22% of total production.

Table 3.5. Sub-sectors in the UK furniture industry, 1997

	Enterprises	Employment ('000)	Production value (ECU m)
Chairs and seats	1,146	38.8	3,349.2
Other office & shop furniture	1,287	29.4	2,724.5
Other kitchen furniture	722	15.7	1,965.0
Other furniture	4,290	45.5	3,423.4
Mattresses	84	8.1	668.2
Total furniture	7,529	137.5	12,130.3

Source: UK Census of Production, 1997.

The furniture industry in Northern Ireland is relatively small. In 1993 output from this region was only 2.1% of the overall UK total while employment (at 3,000) was 2.6% of the UK figure. The majority of firms in Northern Ireland specialise in domestic furniture and serve a predominately local market. However, these companies have the potential to treble the size of their market by expanding into the ROI but here they would undoubtedly face strong competition from both Irish and other British firms where companies such as Habitat, MFI and the DIY multiples (e.g. Texas and B&Q) hold a significant share of the market.

3.1.2 Republic of Ireland

Traditionally small firms operating at a local level have dominated the furniture industry in the ROI. During the 1990s there was modest growth in the sector and by 1997 there were just over 5,600 workers employed (Table 3.6). Much of this growth was as a direct result of demographic changes that have resulted in increased household formation. In addition, the economy of the ROI developed rapidly during the 1990s which resulted in increased living standards and disposable income. The furniture sector benefited from the subsequent growth in consumer expenditure.

The furniture industry is an important purchaser of raw materials from other industries, such as wooden material, plastics, textiles, tanned leather and chemicals.

On average, the production value was twice the value of the purchases of raw materials and other inputs. The main trading partner of the ROI is the UK, 64% of furniture exports were to the UK in 2000. Other important European trading partners were Italy (7% of exports) and Germany (3% of exports).

Table 3.6. Sub-sectors in the ROI furniture industry, 1997

	Enterprises	Employment	Production value (ECU m)
Chairs and seats	32	725	64.0
Other office & shop furniture	54	1,196	84.3
Other kitchen furniture	93	1,043	66.5
Other furniture	112	2,022	143.8
Mattresses	15	641	54.3
Total furniture	306	5,627	412.9

Source: Panorama of EU Business, 2000.

3.1.3 Germany

The German furniture industry is the largest in Europe accounting for almost 30% of the total production value of furniture in the EU. As in the UK/ROI, small firms dominate the sector with a concentration in the "other furniture" and "seating" market (Table 3.7).

Table 3.7. Sub-sectors in the furniture branch of wood manufacturing, Germany, 2000

	Enterprises	Employment	Turnover (€ bn)
Chairs and seats	325	46,720	7.29
Office & shop furniture	289	23,022	2.88
Kitchen furniture	138	22,595	3.91
Other furniture	606	64,003	7.85
Mattresses	61	5,130	0.79
Total furniture	1,419	161,470	22.72

Source: Statistisches Bundesamt, 2001.

There is strong competition in the industry. Labour intensive production has shifted to countries of Eastern Europe, primarily Poland, because of the high labour costs in Germany. There has been a growing trend towards companies producing less on their own but importing furniture to sell along with their own products. This has led to a reduction in employees in this sector since the mid-1990s. Competition has also put prices under pressure and furniture companies are being forced to produce low budget products and reduce production of the more exclusive furniture to maintain market position. Hence, small companies with an exclusive model line and no opportunity to produce cheap products are especially vulnerable. However, a factor favouring SMEs is the demand for an extensive range of furniture for differing lifestyles. The Association of German Wood and Plastic Processing Industry noted in their annual report that there is now no generally accepted furniture style because society has changed and no longer wants to copy the style of their monarchs as in Victorian times.

3.1.4 Italy

The research work concentrated on those firms located within the "Chair Triangle" in North-East Italy, an area that stretches along the eastern strip of the Provence of Udine in the region of Friuli-Venezia Giulia.

The Chair Triangle is one of the most outstanding successes of Italian production serving an international market. There are about 1,200 firms manufacturing chairs but the market for tables and furnishing accessories is expanding. The Friuli chair industry provides employment for about 14,000 people involved in direct manufacture and related services. A further 250 companies, with around 4,000 employees, market finished products.

More than 40 million chairs are manufactured each year, which represents 80% of Italian production, 50% of European production and 33% of the world's seating production. Chairs from Friuli are famous, not for the quantity of production alone, but for their elegant and exclusive designs. The Triangle's production is strongly oriented towards the export market with about 90% of the chairs produced being sold abroad. Countries within the EU import a substantial proportion of these sales but there is also a growing market in the USA and Japan.

The main characteristic of the district is the emphasis that is placed on subdividing the production process. Specialised factories, that are usually quite small, generally carry out each single phase, if not each process of the same phase. This system lends itself to a high degree of specialisation that produces an internationally recognised quality product. This implies a high degree of elasticity in the system, which has the ability to quickly adapt to market demands.

Innovations in the production cycle are continuous, from machines that are often made by local firms to in-house process modifications. Automation is increasingly being adopted in the coating plants and there is a growing trend towards water-based coatings as they are less polluting than those traditionally used. Companies pay attention to customer services particularly regarding flexibility, response to customers' demands and the capacity to guarantee the supply of a range of different products. Some of the leading firms in the industry have also invested in retail stores that sell directly to the consumer.

3.2 Textile Industry

The European textile industry (NACE 17) accounted for about 2.5% of total manufacturing output in 1997. During the 1990s this sector has experienced severe competition from low-wage economies and the EU manufacturers of textiles have responded by switching production to high value goods using modern technologies. Of the sample countries, Italy was the largest producer with high labour productivity and turnover (Table 3.8). Labour productivity has been improving in all sample countries mostly as a result of the contraction in the workforce. The dominant position of Italy was reinforced as the only country with a trading surplus (Table 3.9). As with the furniture industry, the manufacture of textiles is undertaken in predominately small companies. In Italy 78% of enterprises employ fewer than 10 employees, in the UK the figure was 65% although in the ROI only 28% of enterprises employed fewer that 10 employees.[10]

Table 3.8. Economic indicators, textile industry, 1997

	Germany	Italy	UK	ROI	EU
Employment ('000)	129.9	198.3	164.8	8.7	950.0
Turnover (ECU bn)	16.1	28.1	14.1	0.6	95.5
Production value (ECU bn)	14.8	28.2	13.4	0.6	92.2
Value added (ECU bn)	5.1	8.0	5.2	0.2	30.2
Labour productivity ('000 ECU per employee)	39.3	40.3	31.6	23.0	31.8
Turnover per head ('000 ECU)	123.9	141.7	85.6	69.1	100.5

Source: Panorama of EU Industry, 2000.

Table 3.9. Trade balance, textile industry, 1999 (€ million)

	Exports	Imports	Trade balance
Germany	10,480	12,009	-1,529
Italy	13,863	5,872	7,991
UK	4,544	8,109	-3,565
ROI	374	750	-376

Source: Panorama of European Industry, 2000.

The research concentrated on the wet processes in the textile industry namely the preparation and spinning of textile fibres and textile finishing (NACE 17.1 and 17.3 respectively). Within the EU, these two sub-sectors accounted for just over one-quarter of production and almost 30% of employment in the sector as a whole. Output of both the preparation and spinning of textile fibres and finishing of textiles was largest in Italy, followed by Germany (Tables 3.10 and 3.11). In terms of the relative importance of these sub-sectors within each country, Italy dominated with the two sub-sectors combined accounting for 34% of output and 40% of employment of the textile sector as a whole.

[10] No comparable data was available for Germany.

Table 3.10. Economic indicators, preparation and spinning of textiles, 1997

	Production (ECU m)	Employment (000's)	Value added (ECU m)	Labour productivity ('000 ECU per employee)
Germany	2,041	16.0	616	38.5
Italy	5,944	41.2	1,620	39.4
UK	1,964	21.5	654	30.4
ROI	100	1.0	36	34.8
EU	15,908	144.3	4,787	33.2

Source: Panorama of European Industry, 2000.

Table 3.11. Economic indicators, finishing of textiles, 1997

	Production (ECU m)	Employment (000's)	Value added (ECU m)	Labour productivity ('000 ECU per employee)
Germany	1,625	16.4	578	35.3
Italy	3,564	36.6	1,398	38.1
UK	1,094	14.8	511	34.5
ROI	n.a.	n.a	n.a	n.a
EU	11,201	120.8	4,241	35.1

Source: Panorama of European Industry, 2000.

3.2.1 United Kingdom

In 1980 this sector accounted for about 5% of all manufacturing employment but by 1999 employment in the sector had been halved and its share of manufacturing employment had fallen to 3.8%. Falling employment was a common feature of the European textile industry as it struggled to compete with cheap imports from non-EU countries particularly the Far East. UK manufacturers have found it increasingly difficult to compete in world markets. However, the industry has diversified and restructured so that net output has maintained its share of total manufacturing output at around 3%.

Although the UK textile industry includes some very large firms, e.g. Coats Viyella, it is mostly a myriad of SME's which subcontract much of their work to other companies. In 1997, there were just over 6,000 textile establishments, 78% of which employed fewer than 20 employees. These small firms accounted for 15% of the employment and production value of the sector.

This industry is of particular importance in Northern Ireland and consists of a mixture of large multinationals (e.g. Du Pont) and small long established family businesses producing a range of textiles from traditional linen to technical textiles. Employment at around 10,000 (1998 figure) is about 10% of the manufacturing workforce, which is a substantial reduction on the 1980 figure of 26,500 or 19% of all manufacturing jobs. The sector has been heavily dependent on the Irish and

British markets but in recent years has been expanding into other European markets.

3.2.2 Republic of Ireland

Over the last 20 years the Irish textile sector has been decreasing in importance, employment in the sector at 8,000 is half of that in 1980. Net output had been increasing until the mid-1990s when it began a gradual decline, an indication of the declining status of textiles in the Irish economy. There are a substantial number of relatively large firms in the textile industry compared with the overall manufacturing sector. Within the sector, small firms (less than 20 employees) accounted for 48% of enterprises, 15% of employment and 10% of production. Large firms, particularly foreign owned, are important in the yarns and fibres, carpets and other fabric sub-sectors. Investment in textiles in 1995 was 8% of gross turnover (compared with chemicals/pharmaceutical at 5% and food/drink/tobacco at 2%). Much of this investment is concentrated within the yarns and fibres sub-sectors.

3.2.3 Germany

In common with other European countries, Germany has experienced significant contraction in terms of employment and the stock of companies. Employment fell from 326,000 in 1977 to 127,000 by 1998 (a fall of 61%) with a corresponding decrease of 37% in the number of textile companies. Despite this contraction the German textile industry is still one of the major industrial sectors in terms of turnover and employment. Textile production is part of an international process e.g. from a cotton plant in the United States to a spinner and weaver in Italy, a sewing factory in Korea and finally to a finishing company in Germany. Hence, the textile industry is very exposed to global competition and within the last twenty years this process of globalisation has had an intensive impact on the German textile industry. For example, low-grade tasks are outsourced to low wage countries. The most important trading partners for exports are the Netherlands, France, Italy and the UK. The main suppliers are Italy and China but imports from Eastern European countries have been increased.

Clothing accounts for about 45% of textile output, home textiles 30% and technical textiles 25%. Technical textiles for filters or as a substitute for steel used in reinforced concrete are considered to be the growth area of the future. The demand for technical textiles does not depend on fluctuating consumer trends and there are opportunities to expand exports into future growth markets such as China and India.

Future markets are no longer located in Western Europe thus the textile industry has moved a considerable part of its production abroad where market demand is rising and labour costs are low. The shift of production capacity has already reached a high level so that currently there are 63,000 people working for the German textile industry in foreign countries.

3.2.4 Italy

As stated earlier, Italy is the leading European producer of textiles. The research concentrates on textile production in the region Veneto. In Veneto, there were 6,380 firms in the textile and clothing sectors with almost of 90,000 employees and a turnover of around €5.8 bn.[11] The majority of firms are small (fewer than 10 employees) and the firms can be divided into thee types according to the level of organisational complexity and structure. The first group, with about one-third of the firms, represents the more autonomous and structured firms whereby they set their own strategies, outsource production, sell to the market under their own logo or supply clients with a specialised product based on the clients design instructions. The second group comprises about half of the firms and is characterised by the dependence on a single customer. The firms undertake mass production work and have an intermediate organisation complexity. The last group (at 18%) have low organizational complexity, work for other suppliers and undertake only a few stages of the production chain. The main destination of exports is Europe, in particular Germany and France, followed by the Far East, USA and China.

3.3 Fruit and Vegetable Processing

In 1996, food and drinks was the largest industrial sector in the EU accounting for a 15% share of manufacturing activity in the EU in terms of production value, totaling €530 bn. It has relatively low concentration with almost 26,000 companies and is the EU's third largest employer with around 2.6 million employees.

The importance of the EU fruit and vegetable processing and preserving sub-sector (NACE 15.3) is small relative to other sub-sectors in the industry accounting (in 1999) for 5.9% of total output in the food and drinks industry and 0.8% of total manufacturing. The respective employment shares were 6.6% and 0.8%. The production value of the fruit and vegetables industry in the EU was €31.5 bn in 1999, whilst the industry employed 175.6 thousand persons. The EU exported 6.6% of total output (€2.1 bn) but imported goods to the value of €4.8 bn, in 1999.

Output of the EU fruit and vegetables industry showed a large increase in 1991, growing 9.5% in real terms and outperforming the total for food and drinks by 2.4 percentage points. However, after this date output growth slowed so that between 1992 and 1997 the annual average growth rate was equal to just 0.8%, 0.5 percentage points below the average figure for food and drinks activities.

[11] 1993 data, which is the latest available.

Table 3.12. Economic indicators, fruit and vegetables industry, 1997

	Germany	Italy	UK	ROI	EU
Employment ('000)	27.4	21.7	33.6	1.7	167.9
Turnover (ECU bn)	5.9	4.4	5.4	0.21	28.7
Production value (ECU bn)	5.6	4.8	5.2	0.17	30.6
Value added (ECU bn)	1.2	1.0	1.8	0.06	7.3
Labour productivity ('000 ECU per employee)	45.9	44.8	52.7	35.9	43.4
Turnover per head ('000 ECU)	216.0	203.6	159.8	122.4	170.9

Source: Panorama of EU Industry, 2000.

Within the sample countries, Germany had the largest share of EU output at 18% with the UK (17%) and Italy (16%) following closely (Table 3.12). In contrast, the ROI had a very small share of total EU output at only 0.55%. Turnover per head in Germany and Italy was well ahead of the EU average but the UK was the leading country for labour productivity.

Small and medium sized firms make up the majority of enterprises within this sub-sector with nearly all enterprises in each of the countries employing fewer than 250 persons. The industry is fragmented with very small firms (less than 10 employees) accounting for 75% of the Italian firms, in the UK the corresponding figure is 73% and in the ROI 42%.

Italy was the only one of the sample countries to operate a trading surplus (Table 3.13). However, when exports as a proportion of total output are considered the ROI exported the highest proportion (39%) closely followed by Italy (35%), then Germany (22%) and lastly the UK (7%).

Table 3.13. Trade balance, fruit and vegetables industry, 1999 (€ million)

	Exports	Imports	Trade balance
Germany	1,211	3,355	-2,144
Italy	1,675	972	702
UK	363	2,238	-1,874
ROI	66	264	-197

Source: Panorama of European Industry, 2000.

3.3.1 United Kingdom

The fruit and vegetables processing sector represented 17% of total food and drinks production and 6.8% of employment (1997 data) and is characterized by SMEs and few large (over 500 employees) firms. Net output since 1979 has followed a gradual upward trend and employment has remained reasonably stable. Within the UK, this sector is mainly concentrated in England. Producers in England were responsible for 92% of total UK production and employed 90% of the industry's workforce.

On the supply side, considerable reorganisation has taken place during the 1990s. Larger firms have had to reorganise production to take account of changes in demand. Increased concentration in the retail sector has created further pressure for reorganisation within the processing sector and intensive competition between retailers is forcing processors to concentrate more on product diversification.

Processed fruit and vegetables account for a relatively small part of the overall fruit and vegetable market. There is a strong belief amongst UK consumers that canned food is either of poor quality or preserved unnaturally and is unhealthy. This attitude has led to reduced sales but there has been an increase in frozen vegetable sales, especially potato products. In addition, improved storage and global sourcing means that fresh products are available all year round. Fruit and vegetable consumption in the UK is predominantly in fresh form rather than processed form. The total retail market for both fresh and processed fruit and vegetables in the UK, in 1993, was £5.3 bn, with fresh produce accounting for 86%.

In Northern Ireland the fruit and vegetable processing sub-sector, is relatively small, about 6% of total sales in the food and drinks industry. It is a very diverse sub-sector and embraces a wide range of businesses from those involved in grading and packing activities to preparing processed vegetable products such as potato crisps. Employment in the industry has been gradually increasing and in 1998 stood at 1,300. Net output has steadily increased during the 1990s to reach £23m by 1998.

3.3.2 Republic of Ireland

This is a relatively insignificant sector in terms of the Irish economy constituting only 2% of total food industry output and 4.9% of employment, in 1996, a position largely unchanged since 1980.

3.3.3 Germany

In 1997, the fruit and vegetables processing sub-sector accounted for 4.7% of the employment and 5.2% of the production of the total food and drinks industry. This sub-sector has in general tracked the development of the food sector as a whole. Food production is a stable industry, in which concentration is limited, the average number of employees is relatively low and turnover has been gradually increasing.

3.3.4 Italy

As in the other countries, this is a relatively small sector, which accounted for only 8.3% of production and 10.3% of employment of the total food and drinks industry in 1996. Within the sector, fruit and vegetable juices, accounted for 24% of output and 20% of employment with the remaining output and employment coming from fruit and vegetable canning, drying and freezing.

4 Furniture Industry

4.1 Introduction

The postal questionnaire (PQ) was distributed to 538 firms in Germany and 572 firms in the UK/ROI that produced furniture for either the domestic or commercial market. Standard sources of information such as trade and telephone directories, business information publications and the Internet were used to identify the sample. The number of valid replies received was 93 for Germany and 105 for the UK/ROI. In Italy a different approach was used because, after the initial mailing to 100 companies fewer than 10 replied. The Italian researchers opted to administer this short questionnaire by telephone, starting with the larger firms in the Friuli-Venezia Giulia region, to achieve the target of 100 responses.

The selection of SMEs for the in-depth face-to-face interviews (FTF) was made by careful consideration of the type of product manufactured, the size and the extent to which, on the basis of information available, the firm could be categorised as compliance only or beyond compliance in terms of its environmental performance. The products made by these firms were; household furniture (chairs, tables, sofas, beds, kitchen cupboards) and office furniture (desks and chairs) and included flat pack products as well as complete items.

The size distribution for the PQ and FTF samples is shown in Tables 4.1 and 4.2. For the PQ sample the average firm size for the UK/ROI was 88, for Germany 76 and for Italy, where the responses were mostly from small firms, the comparable figure was 13. For the FTF sample, the average employment for UK/ROI was 107, Germany 129 and Italy 41.

Table 4.1. Distribution of firms by size group, postal questionnaire

Size groups	UK/ROI		Germany		Italy	
	Number of firms	% of firms	Number of firms	% of firms	Number of firms	% of firms
1-10	14	13.3	5	5.5	55	55.0
11-25	16	15.2	24	26.4	37	37.0
26-50	21	20.0	30	33.0	5	5.0
51-100	24	22.9	15	16.5	3	3.0
101-250	24	22.9	11	12.1	0	0.0
251-500	6	5.7	6	6.6	0	0.0
Total	105	100.0	91	100.0	100	100.0

Table 4.2. Distribution of firms by size group, face-to-face interviews

Size groups	UK/ROI		Germany		Italy	
	Number of firms	% of firms	Number of firms	% of firms	Number of firms	% of firms
1-10	7	21.9	7	21.9	3	9.1
11-25	2	6.2	6	18.8	13	39.4
26-50	1	3.1	4	12.5	12	36.4
51-100	8	25.0	4	12.5	3	9.1
101-250	10	31.3	6	18.8	1	3.0
251-500	4	12.5	5	15.6	1	3.0
Total	32	100.0	32	100.0	33	100.0

For each country, Table 4.3 shows the summary data for the responses to the PQ regarding the environmental initiatives undertaken. Also shown is the number of firms and the average number of initiatives carried out in each of the environmental performance groups (see Sect. 2.2 for details of these initiatives and performance groups).

Table 4.3. Sample size, number of initiatives undertaken and environmental performance groups, postal questionnaire

	UK/ROI		Germany		Italy	
Sample size	105		93		100	
Total number of initiatives	292		329		388	
Average number of initiatives	2.78		3.54		3.88	
Environmental Performance Groups						
	No. of firms	Av. No. of initia-tives	No. of firms	Av. No. of initia-tives	No. of firms	Av. No. of initia-tives
Compliance only	55	1.09	27	1.26	13	1.62
Compliance plus	39	3.79	53	3.87	84	4.15
Excellence	11	7.64	13	6.92	3	6.00

Overall, SMEs in Italy undertook more initiatives than SMEs in the other countries and there was evidence to support a statistically significant difference between countries at the 1% level. However when the data was disaggregated into environmental performance groups there was no significant difference in the average number of initiatives between countries for firms in the compliance only category. There was a significant difference (at $\alpha = 0.05$) between countries for compliance plus firms but not between UK/ROI and Germany, indicating that compliance plus firms in Germany and the UK/ROI had a similar level of environmental performance. For the excellence performance group, there was a significant difference (at $\alpha = 0.05$) only between the UK/ROI and Italy.

The face-to-face interviews with the firms (FTF) enabled a more in-depth consideration of the extent and quality of environmental initiatives undertaken. Firms were asked detailed questions about eight initiatives that were specific to the furniture industry and for each SME an eco-point score was calculated (see Sect. 2.2 for details). Table 4.4 shows that Germany undertook substantially more initiatives than either of the other countries. There was a statistically significant difference (at $\alpha = 0.01$) in the average number of eco-points between Germany and the other countries but not between the UK/ROI and Italy. Similar results were obtained for the compliance only and compliance plus performance groups.[12]

Table 4.4. Sample size, eco-points and environmental performance groups, FTF interviews

	UK/ROI		Germany		Italy	
Sample size	32		32		33	
Total number of eco-points	201		508		193	
Average number of eco-points	6.28		15.89		5.86	
Environmental Performance Groups						
	No. of firms	Av. No. of eco-points	No. of firms	Av. No. of eco-points	No. of firms	Av. No. of eco-points
Compliance only	17	4.68	9	9.84	32	5.63
Compliance plus	15	8.08	12	14.48	1	13.33
Excellence	–	–	11	22.37	–	–

A priori, there was the expectation that Germany would undertake significantly more initiatives that either the UK/ROI or Italy. This was not evident in the PQ results but using the more comprehensive measure of environmental performance adopted in the FTF analysis, Germany as expected was the best performer in terms of initiatives undertaken.

The respondents to the FTF interview were asked to characterise their profitability and expectations of growth in the next three years. In terms of both profitability and expectations, responses in the UK/ROI were extremely optimistic with 93% operating in profit and 78% forecasting growth in sales. German and Italian replies showed a lower proportion operating in profit (56% and 41% respectively) with 34% and 59% replying that they break-even. The expectation of growth in German and Italian firms was also substantially below the UK/ROI level at 53% and 48% respectively. For both profitability and growth, there was no differentiation between the environmental performance groups in the UK/ROI. In Germany, proportionately more of the compliance only firms indicated break-even profitability along with stagnant or falling expected sales, while compliance plus and excellence firms showed positive economic performance. Since there was only

[12] There was only one compliance plus firm in Italy hence no comparison was made between Italy and the other countries in this category.

one compliance plus firm in the Italian sample, no relationship is shown with environmental performance, except that poorer environmental performance in Italy was associated with reported poorer average economic performance.

4.2 Size Effect

Table 4.5 shows the relationship between firm size and environmental performance, measured as the average number of initiatives per firm for the PQ and average number of eco-points per firm for the FTF data. In terms of the PQ data, neither Italy nor Germany exhibited a relationship between size and the number of initiatives undertaken. For the UK/ROI there was a positive linear relationship between employment and the number of initiatives (correlation coefficient 0.350, significant at $\alpha = 0.01$). This was reinforced by a one-way analysis of variance test for differences in the employment means between the environmental performance groups (significant at $\alpha = 0.01$).

The results from the FTF interviews (Table 4.5) for the UK/ROI also showed a positive linear relationship between the size of the firm and environmental performance ($r = 0.459$, at $\alpha = 0.01$). As expected there was no significant difference in average employment between the compliance only and compliance plus groups because these data have been matched hence the effect of size has been controlled. The German FTF data exhibited a positive linear relationship ($r = 0.530$, at $\alpha = 0.01$) and there was a significant difference (at $\alpha = 0.05$) in the employment averages between the environmental performance groups. This was because the sample of firms interviewed in Germany had not been matched in terms of size across environmental performance groups. There was no relationship between size and environmental performance in the Italian FTF data because of the concentration of firms in the lower end of the size distribution (Table 4.2).

Table 4.5. Average number of initiatives (PQ) and eco-points (FTF) by size group

Size groups	UK/ROI		Germany		Italy	
	PQ	FTF	PQ	FTF	PQ	FTF
1-10	1.64	4.00	4.80	14.19	3.98	5.33
11-25	1.88	2.50	3.75	12.33	3.76	5.48
26-50	2.19	5.00	3.07	14.00	3.40	6.16
51-100	3.29	6.95	4.00	15.33	4.33	6.44
101-250	3.92	6.86	3.45	17.68	–	7.00
251-500	3.33	9.66	4.17	22.34	–	6.00
Total	2.78	6.28	3.54	15.89	3.88	5.86

4.3 Competitive Status

The indicators of competitiveness collected in the postal questionnaire were employment in R&D and destination of sales. In all countries there was a strong positive correlation between the number of full-time employees in R&D and size (UK/ROI $r = 0.429$ at $\alpha = 0.01$; Germany $r = 0.403$ at $\alpha = 0.01$; Italy $r = 0.507$ at $\alpha = 0.01$).[13] It was only in the UK/ROI that a positive relationship between the number of initiatives undertaken and R&D existed ($r = 0.307$ at $\alpha = 0.01$) but this relationship did not hold when firm size was taken into account.

A second measure of competitiveness, based on the destination of sales was derived using these data. For each country, Table 4.6 gives the details of the ranking, the firm distribution and the average number of initiatives.

Table 4.6. Sales ranking, postal questionnaire

Rank	Rule	UK/ROI		Germany		Italy	
		No. of firms	Av. No. of Inits	No. of firms	Av. No. of Inits	No. of firms	Av. No. of Inits
1	No exports, national sales <= 10%	16	2.25	14	3.79	3	3.00
2	No exports, national sales > 10%	27	2.85	17	3.59	2	2.00
3	Exports < 10%	25	2.52	30	3.83	2	3.00
4	Exports 10 - 19%	19	2.53	16	3.63	1	6.00
5	Exports >= 20%	18	3.78	14	3.00	92	3.95

For both the UK/ROI and Germany there was no significant relationship between environmental performance and sales destination ranking. In Italy there was a positive correlation ($r = 0.245$ at $\alpha = 0.01$) between destination of sales and environmental performance. This relationship remained valid whenever firm size was taken into account. Care must be taken in interpreting the Italian result because 92% of the sample was located in the highest export group. The Italian position regarding the destination of sales was very different from that in the other countries. In the UK/ROI average exports were about 11% of sales, in Germany the equivalent figure was 8% while in Italy this average was 60%. This was a reflection of the fact that many of the firms sampled were located in the industrial zone of the province of Udine, which forms part of "The Chair Triangle". Output from

[13] It should be noted that in the postal questionnaire employment and the number engaged in R&D were given in pre-defined size bands. Hence Spearman's rank correlation coefficient has been used for these calculations.

this industrial area is strongly orientated to the export market and many firms located here solely serve the international market. The lack of relationship between the destination of sales and environmental performance in the UK/ROI and Germany can also be seen from the data in Table 4.7. Firms that served an export market were not to be found primarily in the excellence category.

Table 4.7. Competitiveness ranking and environmental performance, postal questionnaire

	All firms	Compliance only	Compliance plus	Excellence
	No. (% of firms)	% of firms	% of firms	% of firms
UK/ROI				
No exports, national sales <= 10%	18 (15.2)	62.5	31.3	6.3
No exports, national sales > 10%	27 (25.7)	48.1	37.0	14.8
Exports < 10%	25 (23.8)	56.0	40.0	4.0
Exports 10% - 19%	19 (18.1)	57.9	31.6	10.5
Exports >= 20%	18 (17.1)	38.9	44.4	16.7
Germany				
No exports, national sales <= 10%	14 (15.4)	21.4	57.1	21.4
No exports, national sales > 10%	17 (18.7)	29.4	58.8	11.8
Exports < 10%	30 (33.0)	23.3	60.0	16.7
Exports 10% - 19%	16 (17.6)	31.3	56.3	12.5
Exports >= 20%	14 (15.4)	35.7	57.1	7.1
Italy				
No exports, national sales <= 10%	3 (3.0)	33.3	33.3	33.3
No exports, national sales > 10%	2 (2.0)	100.0	–	–
Exports < 10%	2 (2.0)	50.0	50.0	–
Exports 10% - 19%	1 (1.0)	–	–	100.0
Exports >= 20%	92 (92.0)	9.8	89.1	1.1

In addition to data on the destinations of sales and employment in R&D the interviews with SMEs collected data on the following explanatory variables – productivity per head, age of machinery, labour growth during the preceding five-year period and staff qualifications. Table 4.8 shows the relationship between these measures of competitiveness and environmental performance. For the British/Irish SMEs significant correlation was observed between environmental performance and both productivity and the age of machinery. The other measures of competitiveness did not show any significant correlation with the environmental performance of the firms. In Germany the only measure to show a significant relationship with environmental performance was exports. Hence few competitive variables were correlated with environmental performance and where they did this was not the case across all samples.

Table 4.8. Correlation between measures of competitiveness and environmental performance, face-to-face interviews

	Expected sign	Correlation coefficient	Level of significance
UK/ROI			
Exports (%)	+	-0.058	ns
Age of Machinery	-	-0.497	0.05
Productivity (turnover/capita)	+	0.558	0.01
R&D (% of total employment)	+	0.011	ns
Growth (employment, 1993-1998)	+	0.112	ns
Qualifications (% of total employment)	+	-0.277	ns
Germany			
Exports (%)	+	0.511[a]	0.01
Age of Machinery	-	-0.228	ns
Productivity (turnover/capita)	+	0.058	ns
R&D (% of total employment)	+	-0.106	ns
Growth (employment, 1993-1998)	+	-0.034	ns
Qualifications (% of total employment)	+	0.197	ns
Italy			
Exports (%)	+	-0.182	ns
Age of Machinery	-	0.206	ns
Productivity (turnover/capita)	+	n.a	
R&D (% of total employment)	+	0.269	ns
Growth (employment, 1993-1998)	+	0.205	ns
Qualifications (% of total employment)	+	-0.128	ns

[a] The percentage of sales exported and the number of employees were significantly correlated thus this statistic has been calculated using partial correlation while controlling for the effect of size.

Table 4.9 shows the average value of the competitiveness measures for the UK/ROI for environmental performance groups.[14] Although there was significant positive correlation between the size of the firm and environmental performance for the sample as a whole (see Sect. 4.2), as expected there was no significant difference in average employment between compliance only and compliance plus groups because these data have been matched hence the effect of size has been controlled for. For the UK/ROI data, employment growth was the only competi-

[14] Italy was excluded from this analysis because, on the basis of the international environmental performance ranking, 32 of the 33 sampled firms were categorised as compliance only.

tiveness variable to show a significant difference on the basis of a Mann-Whitney test.[15]

Table 4.9. Average competitiveness measures by environmental performance groups, UK/ROI, face-to-face interviews

	Compliance only	Compliance plus	All firms
Av. Turnover (£ m)	£7.36m	£17.01m	£11.50m
Av. Productivity (output per head, £)	£68,337	£89,584	£76,993
Av. Level of exports (%)	21	14	18
Av. Employment per firm	100	115	107
Av. Employment growth (%)	6.5	96.5	41.0
Av. Development staff per firm	2.04	4.75	3.26
Av. Qualified staff (% of all staff)	65.6	30.7	45.0
Av. Wages (£ per hour)	£6.63	£6.27	£6.48
Av. Age of machinery (years)	7.83	7.12	7.48
Sample size	17	15	32
Average number of eco-points	4.68	8.08	6.28

Note: The term "staff" refers to all employees.

In summary, there was limited evidence to support the hypothesis that the more competitive the firm the more environmental initiatives it will undertake. In the UK/ROI, at the aggregate level, productivity per head and the average age of machinery were found to have a significant relationship with the increased use of environmental processes. Nevertheless, when comparing the differences between the environmental performance groups neither of these measures showed a significant difference. The German data showed a strong association between environmental performance and the level of exports. The Italian data did not exhibit any evidence to support the hypothesis. It is also worth noting that there were no results that supported the hypothesis that a high level of environmental activity would be a burden for the economic performance of SMEs. Companies that showed a high number of environmental initiatives were no less competitive than the average. They did not suffer from any obvious disadvantage, but they made use of the better image and cost reduction possibilities that compensated for the additional efforts.

[15] It was not possible to carry out a similar analysis for Germany because these data had not been matched by plant size and the latter was significantly correlated with the number of eco-points ($r = 0.53$ at $\alpha = 0.01$).

4.4 Environmental Initiatives

The adoption of the individual environmental initiatives by firms in the postal questionnaire sample showed some variation across country (Fig. 4.1).

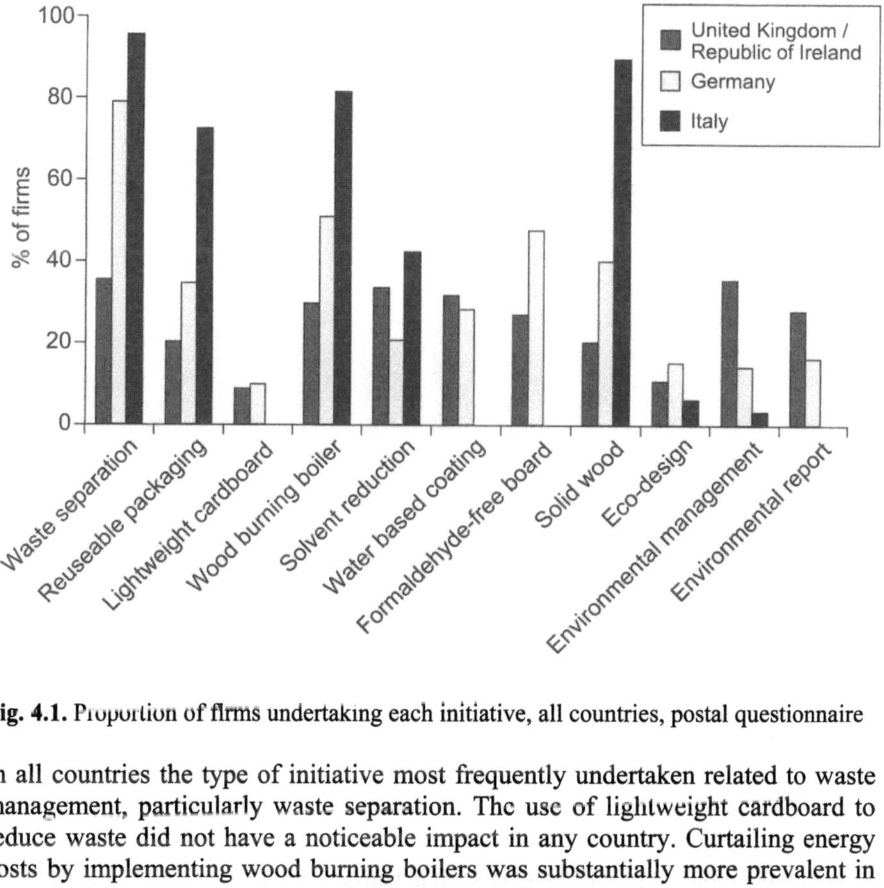

Fig. 4.1. Proportion of firms undertaking each initiative, all countries, postal questionnaire

In all countries the type of initiative most frequently undertaken related to waste management, particularly waste separation. The use of lightweight cardboard to reduce waste did not have a noticeable impact in any country. Curtailing energy costs by implementing wood burning boilers was substantially more prevalent in German and Italian firms than in British/Irish firms. Methods that would lead to reductions in the use of solvents were not as common in the Italian firms with no firms switching to water-based coatings or formaldehyde free boards. The latter was particularly important in Germany. Market focused initiatives, chiefly the use of solid wood were extremely important in Italy and to a lesser extent in Germany and the UK/ROI. Management initiatives that would lead to improved environmental accounting and procedures were virtually absent in Italy but were the second most important category in British/Irish firms.

Not only was there variation in the type of initiative carried out at the macro level but also there was significant variation in the type of initiative undertaken by the three environmental performance groups (Figs. 4.2 to 4.4). Compliance only firms focused on waste separation and undertook virtually no management related activities. Compliance plus firms were involved in waste separation, chemical control and energy efficiency initiatives. Unlike the German and Italian firms, a substantial proportion of compliance plus firms in the UK/ROI undertook management related initiatives. In Germany and the UK/ROI, the excellence firms were distinguished by the almost universal participation in management initiatives and it was only in these firms that eco-design was undertaken with any intensity. Italy showed a similar pattern except that there was only a small proportion of these firms undertaking management initiatives.

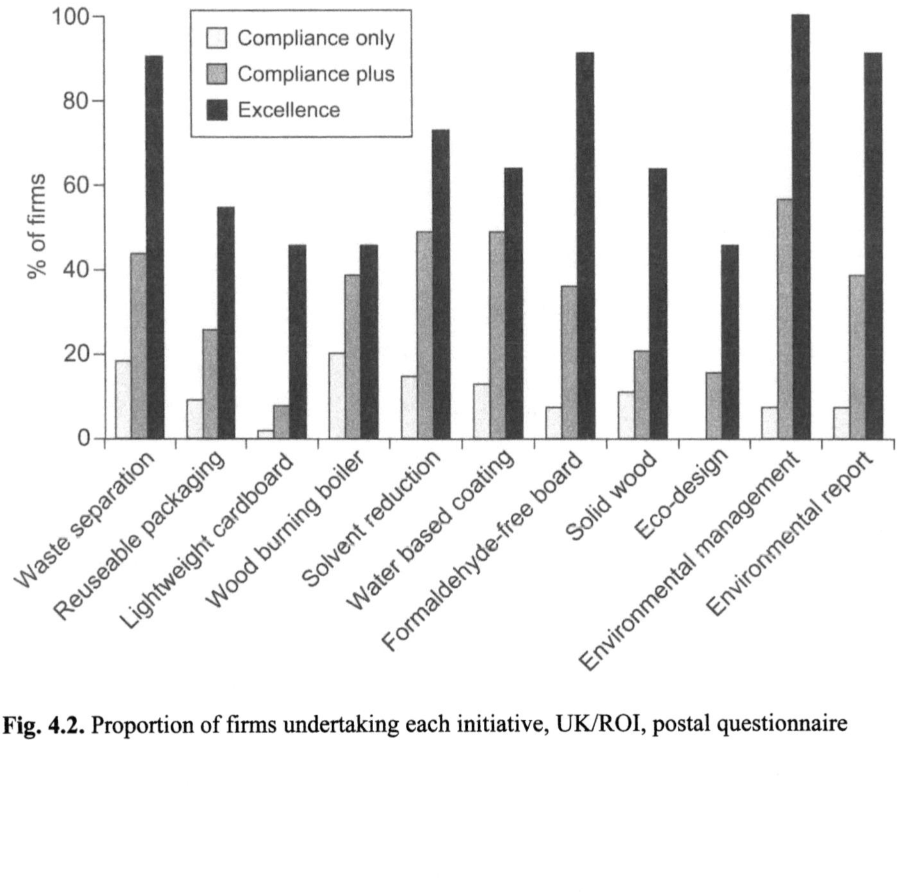

Fig. 4.2. Proportion of firms undertaking each initiative, UK/ROI, postal questionnaire

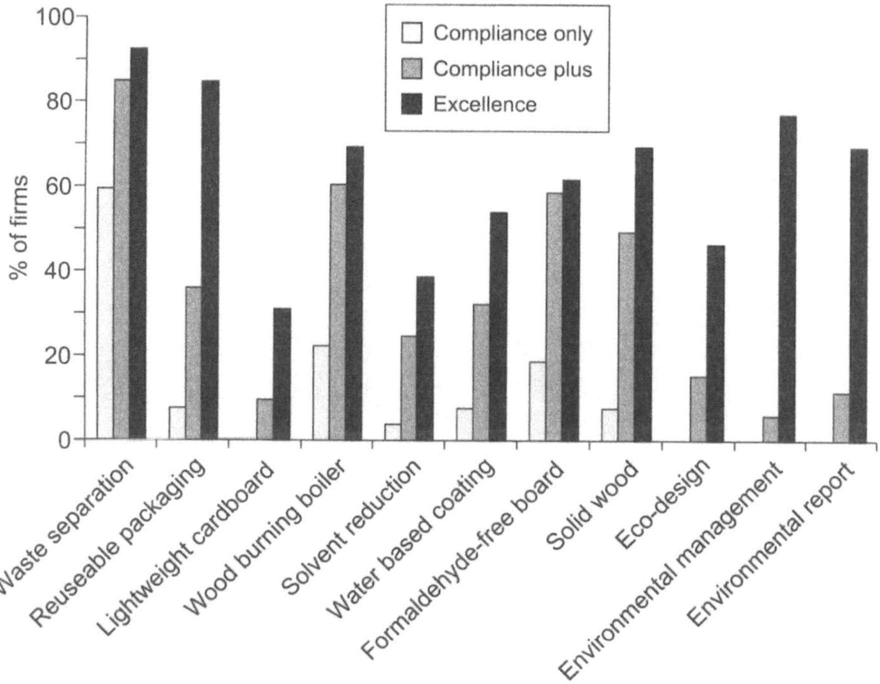

Fig. 4.3. Proportion of firms undertaking each initiative, Germany, postal questionnaire

In general, as firms moved up the hierarchy of environmental performance they advanced from solely "housekeeping" type initiatives to activities that required more technical expertise and finally reached a level where management initiatives were considered to be as important as the basic activities. However, management related initiatives should not be overrated. As one British firm manager described environmental management, it is "more about housekeeping and common sense than major changes in technology".

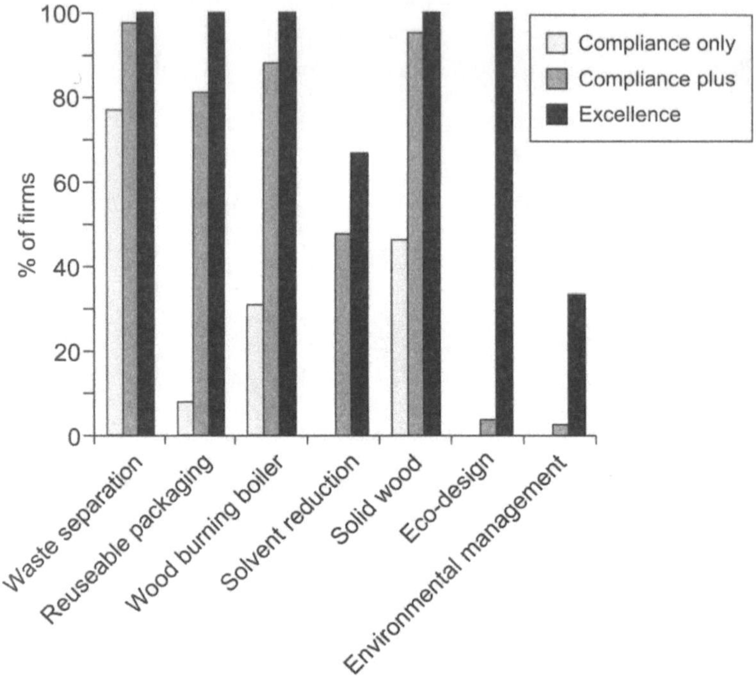

Fig. 4.4. Proportion of firms undertaking each initiative, Italy, postal questionnaire

In the UK/ROI, the results from the face-to-face interviews (Table 4.10) emulated the pattern that was observed in the PQ with waste reduction initiatives receiving almost universal uptake and environmental management initiatives being carried out by a high proportion of firms. However the distinction between the two performance groups was not as marked as with the PQ data and only environmental management showed a significant difference (at $\alpha = 0.01$) between compliance only and compliance plus SMEs. In Germany, there was a more marked differentiation between the environmental performance groups in the uptake of solvent reduction, environmental management and communications, and eco-design (Table 4.11). On the whole, the firms that were classed as excellence had almost universal participation in all the initiatives.

Table 4.10. UK/ROI: Environmental initiatives by performance groups, face-to-face interviews

Initiative	Compliance only		Compliance plus		All firms	
	No. of firms	%	No. of firms	%	No. of firms	%
Waste reduction	17	100.0	14	93.5	31	96.8
Energy efficiency	10	58.8	12	80.0	22	68.7
Reusable packaging	11	64.7	12	80.0	23	71.8
Solvent reduction	12	70.6	9	60.0	21	65.6
Materials change	7	41.2	10	66.7	17	53.1
Environmental management	7	41.2	13	86.7	20	62.5
Eco-design	2	11.8	4	26.7	6	18.7
Sample size	17		15		32	

Table 4.11. Germany: Environmental initiatives by performance groups, face-to-face interviews (figures in brackets are % of sample)

Initiative	Compliance only	Compliance plus	Excellence	All firms
Waste reduction	9 (100.0)	12 (100.0)	11 (100.0)	32 (100.0)
Energy efficiency	5 (55.6)	10 (83.3)	11 (100.0)	26 (81.3)
Reusable packaging	5 (55.6)	8 (66.7)	11 (100.0)	24 (75.0)
Solvent reduction	2 (22.2)	5 (41.7)	10 (90.9)	17 (53.1)
Materials change	8 (88.9)	12 (100.0)	11 (100.0)	31 (96.9)
Environmental management	–	4 (33.3)	9 (81.8)	13 (40.6)
Environmental communication	1 (11.1)	5 (41.7)	11 (100.0)	17 (53.1)
Eco-design	2 (22.2)	4 (33.3)	9 (81.8)	15 (46.9)
Sample size	9	12	11	32

As stated earlier, there was only one firm in the compliance plus category in the Italian sample, thus there is no analysis by performance category. All the Italian firms undertook waste management and packaging initiatives, 94% avoided materials that contained toxic substances and were increasing the use of natural materials. Energy efficiency initiatives were carried out by 70% of firms. A substantially smaller proportion of firms were taking positive steps to reduce the use of solvents (42%) and only about one-third of firms were committed to environmental management systems. These findings endorsed those from the postal questionnaire sample.

While the PQ datasets allow a straightforward enumeration of the type of initiative undertaken using a fairly broad classification, the FTF datasets contain information that enable a more comprehensive analysis. The remainder of this section presents detailed findings for each of these broad categories.[16]

4.4.1 Waste Reduction

Almost every firm in the sample carried out initiatives that were related to waste management. Table 4.12 shows the distribution of firms that separate waste streams (there is no data for the UK/ROI).

Table 4.12. Waste management, Germany and Italy, number of firms, FTF interviews

Waste stream	Germany	Italy
Steel waste	26	4
Other metal waste	15	8
Paper and cardboard	31	32
Plastic foam	14	23
PE-film	23	27
Plastic waste for reuse by suppliers	11	18
Wood	27	28
Special waste	24	18
Household type waste	29	25
Sample size	32	33

As expected, waste management activities relating to paper, wood, PE-film and general waste were commonly undertaken irrespective of country. Steel/metal waste was not separated in many Italian firms, which might reflect the product range of wooden tables and chairs, where higher amounts of steel waste cannot be expected. There was no relationship between the size of the company and the type of waste streams used.

Data relating to the amount of financial investment involved to facilitate waste separation were available only for some of the German sample firms. In general, firms with above average productivity invested more. Typical investments ranged from a waste management register to baling/compressing equipment and wood-burning boilers. The latter also had an impact on reducing energy consumption. There was little evidence of additional investment in labour to support these activities. The German sample also gives an insight into the destination of waste products. For the majority of the firms that provided the information, the data showed that the bulk of waste (over 70%) was either incinerated or went to recycling. Disposal to landfill accounted for a small proportion of waste disposal.

[16] There are constraints with these data in that not all firms provided the information requested and coverage by country varies.

Although there was an absence of quantitative data for the British/Irish firms, there was evidence from the firm interviews that all the firms engaged in some form of waste separation. As in Germany, firms were aspiring to reduce the waste sent to landfill. In the UK the cost of landfill has gradually been increasing over time and firms were very aware of the need to reduce this expenditure. Schemes for reducing or disposing of waste included the use of wood for heating, reworking goods that fail quality tests in order to sell them as seconds, supplying local recycling plants with paper and cardboard, and selling sawdust to other industries. For example, one firm has its sawdust removed by Yorkshire Water who combined it with waste sludge to form a fertilizer product.

In general, the motivation for waste management initiatives has been a combination of regulation and cost. The economic benefits were more pronounced in the British/Irish sample than in the German sample as the former recorded substantial cost savings particularly with regard to the avoidance of landfill costs.

4.4.2 Energy Initiatives

In Germany the main method of achieving energy savings was by using wood or paper waste (chips) to generate heat for the factory and for drying coated furniture (Table 4.13). Italian firms also employed the burning of chips along with the more technically sophisticated system of combined heat and power plants. They also were the greater users of energy efficient lighting.

Table 4.13. Energy initiatives, Germany and Italy, number of firms, FTF interviews

Initiative	Germany	Italy
Energetic use of chips	22	18
Combined heat and power plant	4	20
New energy efficient coating plant	4	1
Energy efficient heating system	7	6
Energy efficient lighting	6	12
Sample size	32	33

Financial investment data in energy saving technology was available only for the German sample. The cost of installing systems to generate energy from chips varied widely, ranging from €6,000 to €1 m. No relationship was found between the level of investment and average productivity. Firms have been installing these systems since the 1980s and although there was no quantitative information on the cost savings generated, the interviews did provide qualitative evidence that this method did lead to appreciable savings. Combined heat and power plants were expensive to install (average in excess of €2 m) and signified the technology of the late 1990s. These plants were found only in companies that employed more than 250 workers. Energy efficient heating and lighting systems were somewhat less expensive to install and where these were in operation they had been installed in the mid-1990s. The former were mainly in firms with more than 100 employees.

Interviews with British/Irish firms document the use of energy saving measures and showed a concentration on wood burning boilers and to a lesser extent, combined heat and power plants. Both of these methods were considered to be expensive, often requiring an investment of up to £500,000 with a payback usually in excess of 5 years. This was sometimes considered to be too long a period by management and hence the system was not implemented. Another factor was size; this form of technology was mostly in firms that employed more than 150 persons. Firms described the housekeeping methods that they employed to reduce energy consumption but these were acknowledged to have little effect in terms of generating significant energy savings.

In all countries, there was a recognition that introducing these initiatives did lead to cost savings and hence increased profits. In addition, the use of waste materials as fuel has helped reduce waste management costs. For example, one British firm said “The company is installing a new wood-burning boiler as the current boiler struggles to meet emission regulations during start-up and shutdown. We aim to maximise its potential by taking wood from other companies for a fee, with a long-term project to consider if electricity generation can be incorporated. Wood-burning will save the company £600 per day on landfill costs, as well as providing free heating for the factory.”

4.4.3 Reusable Packaging

Packing furniture in returnable blankets is the traditional method of distribution to customers. This is both inexpensive and ecologically sound. Table 4.14 shows the various practices employed in the sample countries and shows that Italy did not use returnable blankets as the favoured form of packaging. Italian firms tended to use lightweight cardboard or the environmentally doubtful composite materials for packaging. Both these methods use single way packaging which are most likely a reflection of the comparatively high Italian export rates. Exports do not allow an economically viable method of returning packaging.

Table 4.14. Packaging initiatives, number of firms, FTF interviews

Initiative	UK/ROI	Germany	Italy
Returnable blankets	8	20	11
Cardboard, one way	1	12	2
Cardboard, used several times	1	2	14
Lightweight optimised cardboard	–	2	19
Special returnable packaging	2	3	–
Single material	–	13	–
Composite material	–	3	27
Sample size	32	32	33

In the German and British/Irish firms the main reason for employing reusable packaging was cost and the data showed that cost savings were achieved. In Italy, regulation and market demands were the primary reasons and the resulting economic impact of this initiative was neutral.

4.4.4 Solvent Reduction

The three previous initiatives had a high degree of uptake by SMEs. Of course these initiatives are not specific to the furniture industry, all industrial sectors should be striving to reduce waste, energy usage and packaging. In the furniture industry, glue, stainers, lacquers and varnishes are used extensively. In the past these have been solvent based but there has been a move away from such materials to substitutes that produce lower emissions while maintaining product quality. Legislation, environmental and health and safety, has been the primary motivator for this change. Table 4.15 shows the initiatives associated with solvent reduction.

Table 4.15. Solvent reduction, Germany and Italy, number of firms, FTF interviews

Initiative	Germany	Italy
Solvent free pre-treatment of metals	6	6
Water or powder based coating	7	13
Solvent recovery plant	1	1
Vapour incineration or other end-of-pipe technology	1	2
Hot melt adhesives	2	–
Other (e.g. housekeeping)	7	–
Sample size	32	33

The experience of the German companies has been to concentrate on increasing the use of solvent free treatment of metals and water/powder-based coatings in conjunction with improvements in housekeeping measures. Implementing the first two initiatives required a substantial investment (e.g. the technology to facilitate the use of water-based coatings cost, on average, €750,000) and the resulting cost savings were categorised as being low to medium. Another problem with this technology is that it may initially be difficult to maintain product quality. However, companies that have put this technology into practice reported a reduction in emissions. It was clear from the data that this type of technology was mostly utilized in environmentally excellence firms and to a lesser extent in compliance plus companies. The overall economic impact of solvent reduction technologies has been to improve competitiveness and the firm's image in the market place.

Table 4.16 presents the data on solvent reduction that was available for the UK/ROI.

Table 4.16. Solvent reduction, UK/ROI, FTF interviews

Initiative	No. of firms	Average Investment (£)
UV curing	8	325,000
Water-based staining	12	n.a.
Water-based lacquers	7	n.a.
Compliant glues	9	5,000
HVLP Spray gun (high volume/low pressure)	10	1,800
Compliant lacquers	8	n.a.
Sample size	32	

In contrast to Germany, it was the compliance only firms that appeared to be out performing their compliance plus counterparts with respect to solvent reduction (see Table 4.10). This was certainly true for inexpensive technologies such as HVLP spray guns and compliant lacquers but for the more expensive UV curing technology, compliance plus firms dominated (7 out of the 8 firms). Also, it was the larger firms (over 100 employees) that tended to use UV curing. Many of the solvent reduction initiatives undertaken were made in consultation with suppliers who have positively addressed issues such as the need to maintain final product quality and acceptable lacquer drying time. As one British company stated "Initial attempts to introduce low-VOC lacquers were abandoned after quality problems could not be resolved. The supplier then developed a quicker drying product that proved to be satisfactory."

4.4.5 Avoidance of Toxic Substances in Raw Materials

Data on the avoidance of toxic substances in raw materials was available only for Germany and Italy. The situation was very different in the two countries (Table 4.17). In Germany, over 80% of the companies interviewed refrained from using chipboard that contained formaldehyde. Reduction in the usage of other harmful inputs such as heavy metals and plastics was mostly confined to compliance plus and excellence firms. Real wood was undoubtedly the most frequently used natural material in both countries. However, almost 60% of the German companies sourced wood from sustainable managed forests compared to 6% of the Italian firms. There was also a distinction within the German sample in terms of which companies were concerned about the origin of wood supplies. The tendency was for excellence and compliance plus companies to source from sustainable managed forests with few compliance only companies paying attention to the origin of the material. In general, the German firms were more aware of the problems associated with the different raw materials and were more disposed to substituting these for ecologically sound inputs. Demands from the market (customers) were instrumental in driving this change.

Table 4.17. Toxic reduction and material change in inputs, Germany and Italy, number of firms, FTF interviews

	Germany	Italy
Avoidance of toxic substances in raw materials		
Heavy metals	15	2
Compounds containing chlorine, e.g. PVC	17	3
Formaldehyde e.g. in chipboard	26	2
Pesticides e.g. in wool	14	2
Other	4	–
Use of natural/ecological materials		
Real wood	27	31
Of which wood from sustainable managed forests	16	2
Leather tanned without chromium	8	–
Textiles with eco-labels	9	–
Sample size	32	33

British/Irish firms were also receptive to change but on a much smaller scale. Table 4.10 shows that only 53% of companies participated in switching inputs to materials that were more environmentally friendly. Interviews with these companies recorded almost universal acknowledgement of a move away from hardwood, increased sourcing from sustainable managed forests and, in some cases, steps to reduce material toxicity. As in Germany, the market was cited as the reason for implementing this change.

4.4.6 Eco-Design

Eco-design encompassed activities in which the product was designed in a way that was considered to be environmentally friendly. The most common of the activities undertaken was the construction of durable products using ecologically proven materials, take-back and repair of furniture and the development of timeless designs. In general, established companies were disinclined to enter into a complete redesign of a successful product but were approaching eco-design where they planned and designed new product lines. This sometimes had a knock-on effect on other initiatives e.g. applying environmentally friendly varnishes (natural waxes and oils). Newer companies were pro-active in product design so that they could, for example, make optimal use of wood available from local sources or maximize usage of certain techniques so that the application of solvents was reduced.

Fifteen companies in Germany undertook eco-design, six in the UK/ROI but it was completely absent in the Italian firms sampled. Eleven of the 15 companies in Germany had implemented a formal eco-design scheme. The implementation of eco-design was most frequent amongst the excellence performance group; was market driven and there was no relationship between uptake and the number of employees.

4.4.7 Environmental Communication

Germany was the only country to provide data on environmental communication that included activities such as providing environmental product information, guided visitor tours, production of environmental reports and statements. Environmental communication was undertaken by a total of 17 companies, which included all the excellence firms and 42% of the compliance plus firms (see Table 4.11). There was a significant relationship with size, in that 70% of the firms that carried out this initiative employed more than 100 staff. Although there was no data for the British/Irish firms interviewed, the postal questionnaire showed that 29 firms issued an environmental report or statement and almost all the excellence performing firms were involved in this activity (see Fig. 4.2). Unlike the situation in Germany, there was no relationship between this initiative and numbers employed.

4.4.8 Environmental Management Systems

Of all the initiatives considered in this study, the initiative that most strongly discriminated between the environmental performance groups of compliance only, compliance plus and excellence was that relating to environmental management systems. Using the PQ data and allowing for the fact that firms with EMS will automatically have a higher than average number of initiatives, the difference between firms with EMS and those without EMS was apparent (Table 4.18). In the UK/ROI there was, as expected, a significant positive linear relationship between firm size and the adoption of EMS ($r = 0.328$ at $\alpha = 0.01$). As size increases proportionately more firms undertake management initiatives. The data suggested that there was a threshold employment value of 50, only 18% of firms with under 50 employees implemented management initiatives compared with 55% of firms with employment greater than 50. The PQ data from Italy and Germany did not exhibit any relationship with firm size.

Table 4.18. Average number of initiatives adopted by firms with and without an EMS, postal questionnaire

	Firms with EMS	Firms without EMS
	Average number of initiatives	Average number of initiatives
UK/ROI	4.86 (37)	1.65 (68)
Germany	6.46 (13)	3.06 (80)
Italy	5.33 (3)	3.84 (97)

Note: Figures in parenthesis denote number of firms.

The face-to-face firm interviews provided more extensive information about environmental management systems and related tools. Table 4.19 shows that SMEs that implemented environmental management techniques did, on the whole, undertake more initiatives (even after allowing for the additionality effect). This was particularly the case in the UK/ROI and to a lesser extent true in Germany.[17]

Table 4.19. Average number of initiatives adopted by firms with and without an EMS, across performance group, face-to-face interviews

	Firms with EMS	Firms without EMS
	Average number of eco-points	Average number of eco-points
UK/ROI		
Compliance	7.52 (7)	2.70 (10)
Compliance plus	9.02 (13)	2.00 (2)
Germany		
Compliance	–	9.84 (9)
Compliance plus	15.42 (4)	14.00 (8)
Excellence	23.34 (9)	18.00 (2)
Italy		
All firms	4.80 (12)	2.86 (21)

Note: Figures in parenthesis denote number of firms.

The study was interested in not solely establishing if a firm had an accredited EMS system but whether the firm used management techniques. This would indicate an awareness of the issues involved and the need for quality management tools such as data collection and supplier evaluation. The results for the UK/ROI and Germany are given in Tables 4.20 and 4.21.

Table 4.20. Environmental management tools by size group, UK/ROI, FTF interviews

Size	EP	EMS	EMS Accredited	IP	ET	DC	SE	Total firms
1-10	–	–	–	–	–	–	–	7
11-25	–	–	–	–	–	–	–	2
26-50	–	–	–	–	–	–	–	1
51-100	5	3	2	4	4	3	3	8
101-250	3	5	3	3	3	5	2	10
251-500	3	3	2	2	1	3	2	4
Total	12	11	7	10	8	12	7	32

EP Environmental policy, *EMS* Environmental management system, *IP* Improvement programme, *ET* Environmental team, *DC* Data collection, *SE* Supplier evaluation.

[17] Italy had only one compliance plus firm hence the aggregate figure is presented.

Table 4.21. Environmental management tools by size group, Germany, FTF interviews

Size	EP	EMS	EMS Accredited	IP	ET	DC	SE	Total firms
1-10	–	–	–	–	–	–	6	7
11-25	–	–	–	–	–	–	4	6
26-50	–	–	–	–	–	–	2	4
51-100	1	2	2	3	1	1	2	4
101-250	3	3	3	4	5	2	5	6
251-500	3	2	5	4	4	3	3	5
Total	7	7	10	11	10	6	22	32

EP Environmental policy, *EMS* Environmental management system, *IP* Improvement programme, *ET* Environmental team, *DC* Data collection, *SE* Supplier evaluation.

In both Germany and the UK/ROI there was an association with firm size. The sample results showed that firms with fewer than 50 employees did not participate in environmental management related initiatives. There was an exception to this in that some small German companies were involved in evaluating the environmental performance of suppliers as part of the 'normal' management routine. In addition to the correlation with firm size there was also a positive relationship between environmental management initiatives and productivity (measured as turnover per capita).

The German sample provided limited data on the initial set-up costs of an EMS. The median cost was €30,000 with an average payback period of 2-3 years. The motivation for firms in the UK/ROI and Germany to investigate environmental management initiatives was primarily the market, particularly demands from corporate customers, government contracts and venture capitalists.

4.5 Drivers for Undertaking Initiatives

Tables 4.22 (PQ) and 4.23 (FTF) give the breakdown of the dominant drivers for each initiative undertaken. The drivers considered were regulation, cost, market (customer and supplier demands) and health and safety (H&S). In some instances not all the firms surveyed responded to this question hence the data does not sum to 100 while in other cases firms gave more than one response thus the percentages may sum to over 100.

Table 4.22. Dominant drivers (cited by > 20% of firms), postal questionnaire

Initiative	UK/ROI	Germany	Italy
Waste separation	Regulation 68% Cost 27%	Regulation 59% Cost 37%	Regulation 98%
Reusable packaging	Cost 81%	Cost 63% Market 28%	Cost 50% Regulation 50%
Lightweight cardboard	Cost 50%	Cost 78%	–
Wood burning boiler	Cost 61% Regulation 32%	Cost 87% Regulation 23%	Cost 100%
Solvent reduction	H&S 66% Regulation 60%	H&S 47% Regulation 42%	Regulation 71% H&S 29%
Water based coating	H&S 52% Regulation 48%	H&S 46% Market 31%	–
Formaldehyde-free board	H&S 68%	Regulation 59% Market 57%	–
Solid wood	Market 76%	Market 82% H&S 26%	Market 75% H&S 25%
Eco-design	Market 27%	Market 100%	Market 100%
Environmental management	Regulation 65% H&S 49%	Market 62% H&S 38%	Market 100%
Environmental report	Regulation 59% H&S 41%	Regulation 53% Market 47%	–

Table 4.23. Dominant drivers (cited by > 20% of firms), face-to-face interviews

Initiative	UK/ROI	Germany	Italy
Waste reduction	Cost 65% Regulation 33%	Regulation 50% Cost 28%	Regulation 85%
Energy efficiency	Cost 82%	Cost 73%	Cost 87%
Reusable packaging	Cost 74%	Cost 43% Market 23%	Regulation 24% Cost 12%
Solvent reduction	Regulation 62% H&S 19%	Market 33% H&S 22%	Regulation 57% H&S 29%
Materials change	Market 54%	Market 74%	–
Environmental management	Market 55% Regulation 20%	Market 69%	H&S 83%
Environmental communication	–	Market 65%	–
Eco-design	Market 100%	Market 85%	–

On the whole the tables show a reasonable degree of agreement between the two surveys although in Germany the FTF results indicated a greater emphasis on customer and supplier demands for solvent reduction initiatives. In Italy and to some extent in the UK/ROI, there was a discrepancy in the main driver for management initiatives. There was little variation between countries. Overall, waste management initiatives were mainly driven by regulation and cost (of disposal). Packaging initiatives were mostly cost driven except in Italy where packaging legislation had an impact. Energy initiatives were almost entirely cost driven although the PQ data did show that the move to wood burning boilers was also related to a need to reduce emissions in order to comply with legislation. For solvent reduction initiatives, if H&S reasons were considered as a form of regulation then the stringency of legislation had a major impact here. The market was the main driver for eco-design, avoidance of toxic substances in raw materials and management initiatives.

In summary, cost was the main driver for the fundamental housekeeping type of initiatives. As the initiatives become more technically demanding, regulation and the market become the key drivers. Finally, for initiatives that were related to design, environmentally friendly inputs and management systems the market was the dominant driver.

4.6 Economic Impact of Initiatives

At the face-to-face interviews respondents were asked to assess the implications of undertaking each of the initiatives in terms of positive, negative or neutral effects on labour, sales, costs, productivity, profits, price, market position and image. The dataset was constrained in that for the UK/ROI there were no indicators of change for profits, price, market position and image and in the Italian dataset there was no data for price or image.

At the aggregate level, using all the available data, there were 144 effects arising from 140 initiatives in the UK/ROI, 240 from 172 initiatives in Germany and 88 from 146 initiatives in Italy. These gave a ratio of effects per initiative of 1.03, 1.40 and 0.60 respectively. For a few of the initiatives there was a certain degree of uniformity in terms of effects, e.g. energy initiatives invariably resulted in reduced costs in all the countries. However, there was also evidence of variability in effect for the same initiative undertaken in the different countries e.g. waste reduction in Germany, solvent reduction in the UK/ROI.

Table 4.24 shows the positive or negative economic impact of each initiative undertaken in terms of sales, labour, costs and productivity in each country. From the firm's position the overall impact of increased sales and reduced costs with associated increased productivity was positive. Cost driven initiatives such as waste reduction, energy efficiency and reusable packaging had the greatest impact on the firm particularly in the UK/ROI. Market driven initiatives such as eco-design and material change did lead to increased sales and productivity. The impact of environmental management initiatives was small although environmental communication initiatives showed an improvement in firm productivity (Germany only). There was little increase in labour associated with the initiatives. In Germany net labour increase was associated with most of the initiatives while in the UK/ROI it was the management initiatives that exhibited the greatest net gains.

From the German dataset, an analysis of the information relating to profits, market position and image, showed an increase overall; with profits +17; position

+24; image +40. The increase in profits was mostly due to energy initiatives while activities associated with environmental communication led to a definite improvement in market position and company image.

Table 4.24. Economic impact, face-to-face interviews

Initiative	Sales		Labour		Costs		Productivity	
	↑	↓	↑	↓	↑	↓	↑	↓
Waste reduction								
UK/ROI			1		5	20		
Germany		2			7	7	3	2
Italy					1			
Energy efficiency								
UK/ROI	1					20	1	
Germany	3		2			9	5	
Italy		6			1	9		5
Reusable packaging								
UK/ROI	1		1		1	21	2	
Germany	2		2		1	4	3	
Italy	7				2	5		5
Solvent reduction								
UK/ROI	2		3	3	6	8	8	
Germany	2		1		1	4	7	
Italy	1	1				1	2	
Materials change								
UK/ROI	6				1			
Germany	3		1		2		5	
Italy					2			2
Env. management								
UK/ROI	4		4		4	4	3	
Germany	1		1			1	4	
Italy								
Env. communication								
UK/ROI								
Germany	3		2		3	1	8	
Italy								
Eco-design								
UK/ROI	6			1		4	3	
Germany	3	1	3		2	1	6	
Italy								
Total impact								
UK/ROI	20	0	9	4	17	77	17	0
Germany	17	3	12	0	16	27	41	2
Italy	8	7	0	0	6	15	2	12

Table 4.25. Economic impact, UK/ROI and Germany, postal questionnaire

Initiative	Sales		Labour		Costs		Profits	
	↑	↓	↑	↓	↑	↓	↑	↓
Waste separation								
UK/ROI			6	1	9	13	8	4
Germany	2	2	34	1	29	15	4	17
Reusable packaging								
UK/ROI	1	2	3	3	5	6	7	1
Germany	7	1	7	6	9	16	6	7
Lightweight cardboard								
UK/ROI				1	2	2	1	
Germany			2	1	4	4	4	3
Wood burning boilers								
UK/ROI			6	3	4	14	9	1
Germany		3	8	5	8	27	16	7
Solvent reduction								
UK/ROI	2	1	5	2	12	6	1	8
Germany	3		6		5	3	3	3
Water based coating								
UK/ROI	7	3	2	4	13	5	6	4
Germany	9		8	3	12	4	6	11
Formaldehyde free chipboard								
UK/ROI	2		4	1	9			2
Germany	15				20	2	3	11
Solid wood								
UK/ROI	6		6		10	2	8	2
Germany	18	1	10	1	21		11	6
Eco-design								
UK/ROI	1			1	3	2	2	
Germany	8		6		5		3	2
Env. management								
UK/ROI	1	2	4		15		2	6
Germany	4		5		7		1	3
Env. peport								
UK/ROI		1	3	1	9	1	2	4
Germany			6		7			4
Total Impact								
UK/ROI	20	9	35	17	91	51	46	32
Germany	66	7	96	17	127	71	57	74

Responses to the PQ regarding the economic impact on the firm of initiatives undertaken are presented in Table 4.25. There was no comparable data for Italy hence it has been excluded from the analysis. In the UK/ROI there were 301 total effects from 292 initiatives giving a ratio of 1.03 effects per initiative. The comparable figures for Germany were 515 effects from 329 initiatives, a ratio of 1.56 effects per initiative. Thus both the FTF and PQ show that Germany experienced the highest ratio of effects per initiative. As before the data exhibited variability in effects for the same initiative. For example, for the waste separation initiative only 40% of the British/Irish firms that recorded an effect indicated that this initiative resulted in increased costs while in Germany 65% of the firms that noted an effect experienced increased costs. However, this unevenness in effects was not applicable to all the initiatives undertaken. Management related initiatives did show almost uniformity in effects with almost all the firms showing evidence of increased costs and labour along with reduced profitability.

Whereas the experience of the firms that were interviewed (i.e. FTF) was reasonably positive in terms of falling costs, increasing sales and hence profitability. The situation for the firms that returned the PQ was somewhat different. The overall net effect for these firms was negative particularly regarding costs and profits. In the UK/ROI cost driven initiatives such as the installation of wood burning boilers and waste management did lead to small cost savings. However, these were counteracted by increased costs for initiatives, such as solvent/chemical reduction and management, which were driven primarily by legislation (environmental and H&S).

In Germany the discrepancy between PQ and FTF was even more apparent with reusable packaging and wood burning boilers the only initiatives to show a net fall in costs. Initiatives that had a strong market driver (solid wood, eco-design, formaldehyde-free materials) accounted for about 60% of the increased sales but these were generally offset by increased costs hence there was little improvement in profits.

The total impact on labour was substantially greater in Germany than in the UK/ROI. Waste separation, the use of solid wood and management related initiatives were the main contributors to this increase in labour in Germany and in the UK/ROI.

One strand of the research was to investigate the relationship between the environmental performance of the firm, when ranked as being compliance only, compliance plus and excellence, and the associated economic impact that undertaking these environmental initiatives had on the firm. Constraints within the datasets allowed this analysis to be carried out only with data from the postal survey and also excluded Italy because of the incompleteness of the Italian data. When the responses from the PQ for the UK/ROI and Germany were combined, the distribution of economic effects, the total number of initiatives undertaken and the effects to initiative ratio for each environmental performance group is given in Table 4.26.

Table 4.26. Economic impact, performance groups and size, UK/ROI and Germany, PQ

	Sales		Labour		Costs		Profits		TE	TI	E/I
	↑	↓	↑	↓	↑	↓	↑	↓			
C - only	14	2	31	5	31	22	14	19	138	94	1.47
C - plus	48	9	67	25	123	65	63	58	458	353	1.30
Ex.	24	5	33	4	64	35	26	29	220	174	1.26
1-25	21	3	34	1	72	23	27	41	222	167	1.33
26-50	27	3	38	9	52	31	28	26	214	138	1.55
51-100	15	1	26	9	33	30	16	8	138	139	0.99
> 100	23	9	33	15	61	38	32	31	242	177	1.37
Total	86	16	131	34	218	122	103	106	816	621	1.31

C - only Compliance only, *C - plus* Compliance plus, *Ex.* Excellence, *TE* Total effects, *TI* Total initiatives, *E/I* Effects per initiatives.

The final column shows that the ratio of effects to initiatives undertaken decreases as environmental performance increases. This was because as environmental performance improves, firms move along a chain from undertaking solely basic housekeeping type initiatives, for example, waste management to incorporating technical and management related activities in the business. As firms undertake more technical and management functions the economic effects of the associated environmental initiatives were less obvious and firms in this sector found it more difficult to quantify the outcomes.

From the firm's perspective a positive economic effect would be increased sales, reduced costs and consequently rising profits. If the ratio of effects to initiatives is calculated to include only these factors then the ratio for compliance only firms would be 0.53; for compliance plus the ratio would be 0.50 and 0.49 for excellence firms. This would indicate that compliance only firms were able to achieve similar economic benefits as the better environmental performing firms.

A breakdown of the contribution of labour for the three performance groups shows that among the compliance only firms, 58% of the increase in labour can be attributed to waste management and energy initiatives. This was in contrast to 36% in the excellence firms. In the latter group firms have returned a similar proportion pertaining to management related initiatives.

Table 4.26 also provides results on the effects and initiatives for the different employment size bands. At the aggregate level there does not appear to be a relationship between effects and size. However, if the positive economic effects only are considered then the very small firms did seem to be at a cost disadvantage. These firms have substantially higher costs in undertaking environmental initiatives than medium to large sized firms. Some of this increased cost could be attributed to the increases in labour that were required to undertake the environmental initiatives.

4.7 Obstacles to Adoption of Environmental Initiatives

In the course of the face-to-face interviews the managers were asked to rank in order of importance the factors that impeded the introduction of environmental initiatives into the organisation. Table 4.27 shows the ranking for each of these factors by country. The ranking is based on the frequency distribution of answers with a "1" indicating the most frequently cited problem, a "2" the next most important obstacle and so on.

Table 4.27. Ranking of obstacles to adoption, face-to-face interviews

Problem	UK/ROI	Germany	Italy
Difficult to raise required capital	2	1	3
Payback period on clean technology is too long	4	4	–
Shortage of required skills and expertise	4	4	4
Clean technology is risky and unproven	–	3	–
Management priorities are focused on other issues	2	6	2
Lack of management time	3	5	–
Profit motive supersedes environmental initiatives	5	–	5
Lack of good quality advice	1	6	5
Lack of help from suppliers	4	6	–
Environmental consultancy services are expensive	–	–	3
Regulations are too uncertain to plan for new technology	1	2	1

There was reasonable agreement between the countries that the most important constraints on the implementation of environmental initiatives into businesses were (i) the difficulties faced in raising capital for a new investment and (ii) that the regulations were too uncertain to allow managers the scope to plan for putting new technologies into practice. The latter was a reflection of a general shortfall in both knowledge about new technologies and the availability of information concerning current or proposed legislation. The British/Irish managers reinforced this perception by frequently citing the shortage of good quality advice. The opposite was the case for the Italian and German companies. Here the matter of receiving good quality advice was not considered to be a major obstacle although for the Italian firms specific, targeted advice from consultancy services was expensive.

Italian and British/Irish firm managers also considered that from a management point of view there were too many other priorities and hence environmental issues came low on a list of concerns. In addition, the British/Irish firms believed that the shortage of management time to address such issues was an important constraint.

Table 4.28 shows a breakdown of these obstacles by environmental performance group. It is interesting that in Germany the top three obstacles were ranked in the same order irrespective of performance group. These were (i) problems in finding the necessary capital, (ii) the uncertainty of regulation and (iii) the unproven nature of clean technology. In the UK/ROI compliance only firms were mostly worried about acquiring the necessary capital, the imperfect nature of advice, demands on management time and business concerns. In contrast, compliance plus firms considered the regulation and information issues to be the most important. There was no association between the size of the firm and the ranking of the obstacles, small and large firms alike endorsed the above findings.

Table 4.28. Ranking of obstacles to adoption by environmental performance groups, UK/ROI and Germany, face-to-face interviews

Problem	UK/ROI		Germany		
	C	C+	C	C+	E
Difficult to raise required capital	1	4	1	1	1
Payback period on clean technology is too long	4	6	4	6	4
Shortage of required skills and expertise	4	6	3	5	5
Clean technology is risky and unproven	–	–	3	3	3
Management priorities are focused on other issues	2	3	–	–	–
Lack of management time	2	5	–	4	6
Profit motive supersedes environmental initiatives	2	–	–	–	–
Lack of good quality advice	1	2	–	6	–
Lack of help from suppliers	3	5	–	6	–
Environmental consultancy services are expensive	–	–	–	–	–
Regulations are too uncertain to plan for new technology	3	1	2	2	2

C Compliance only, *C+* Compliance plus, *E* Excellence.

4.8 Environmental Strategy

The extent of implementation of clean technology in the manufacturing process in the firm may be guided by the company's environmental strategy. In all the countries, the majority of firms that responded to the PQ viewed complying with regulations as the foremost focus of this strategy (Table 4.29). However, the actual proportion of firms adopting this regulation strategy varied between countries. As shown in the Table, Italian firms almost universally adopted this regulation strategy whereas in the UK/ROI about one-quarter of the firms engaged in a proactive environmental strategy through eco-practices. Germany showed the most evidence of a move away from the reactive regulation strategy to embrace a greater mix of strategies that included focusing on inducing cost savings and positive market strategies that affected both practices and products.

In terms of differences in company environmental strategies between the three performance groups, the Italian data showed no differences i.e. companies adopted a regulation strategy irrespective of environmental performance. In the UK/ROI the only strategy to exhibit a significant difference (at $\alpha = 0.10$) between the performance groups was that of promoting eco-products with the excellence firms surpassing the compliance only and compliance plus firms. In Germany there was a significant difference (at $\alpha = 0.10$) for both regulation and eco-practice strategies. As firms move beyond compliance the favoured strategy moves away from being regulation focused to the market-centred strategy of improving brand image by undertaking environmentally friendly practices.

Table 4.29. Firm environmental strategy, number and percentage (in brackets) of firms, postal questionnaire

Main focus	Compliance only	Compliance plus	Excellence	All firms
UK/ROI				
Regulation	45 (81.8)	27 (69.2)	9 (81.8)	81 (77.1)
Cost savings	5 (9.1)	8 (20.5)	2 (18.2)	15 (14.3)
Promoting brand by eco-practice	11 (20.0)	12 (30.8)	4 (36.4)	27 (25.7)
New markets for eco-products	–	1 (2.6)	–	1 (0.9)
Promoting eco-products	3 (5.5)	5 (12.8)	3 (27.3)	11 (10.5)
Total firms	55	39	11	105
Germany				
Regulation	17 (63.0)	27 (50.9)	3 (23.1)	47 (50.5)
Cost savings	13 (48.1)	23 (43.4)	5 (38.5)	41 (44.1)
Promoting brand by eco-practice	4 (14.8)	15 (28.3)	6 (46.2)	25 (26.9)
New markets for eco-products	2 (7.4)	7 (13.2)	1 (7.7)	10 (10.8)
Promoting eco-products	5 (18.5)	17 (32.1)	5 (38.5)	27 (29.0)
Total firms	27	53	13	93
Italy				
Regulation	10 (76.9)	79 (94.0)	2 (66.7)	91 (91.0)
Cost savings	–	2 (2.4)	1 (33.3)	3 (3.0)
Promoting brand by eco-practice	–	1 (1.2)	–	1 (1.0)
New markets for eco-products	–	1 (1.2)	–	1 (1.0)
Promoting eco-products	1 (7.7)	–	–	1 (1.0)
Total firms	13	84	3	100

More than one answer was possible to this question hence the sum of the answers may exceed the total number of firms (and 100%).

Table 4.30. Influences on final decision regarding environmental issues, number and percentage (in brackets) of firms, postal questionnaire

Main influence	Compliance only	Compliance plus	Excellence	All firms
UK/ROI				
Regulation	27 (50.0)	18 (47.4)	6 (60.0)	51 (50.0)
Management awareness	12 (22.2)	10 (26.3)	3 (30.0)	25 (24.5)
Cost	12 (12.2)	3 (7.9)	–	15 (14.7)
Customer demands	3 (5.6)	7 (18.4)	1 (10.0)	11 (10.8)
Total firms	54	38	10	102
Germany				
Regulation	10 (45.5)	21 (42.8)	2 (15.4)	33 (39.3)
Management awareness	3 (13.6)	17 (34.7)	6 (46.2)	26 (31.0)
Cost	–	–	–	–
Customer demands	9 (40.9)	11 (22.5)	5 (38.4)	25 (29.7)
Total firms	22	49	13	84
Italy				
Regulation	12 (92.3)	77 (91.7)	2 (66.7)	91 (91.0)
Management awareness	–	4 (4.8)	1 (33.3)	5 (5.0)
Cost	–	1 (1.2)	–	1 (1.0)
Customer demands	1 (7.7)	2 (2.4)	–	3 (3.0)
Total firms	13	84	3	100

Table 4.30 shows the breakdown of answers to a complementary question, which centred on the factor that most influenced the final decision on environmental issues. As before the Italian results showed a total focus on regulation whereas in both the UK/ROI and Germany the picture was more mixed. Regulation was still important but to a lesser extent. Management awareness accounted for about one-quarter of the British/Irish responses and for almost one-third of the German replies. These answers show that when it comes to making a final decision there is a need for management to be well informed about the possible environmental impact.

Unlike the answers to the previous question on company strategy, there was no difference between the three performance groups regarding the factors that influence the final decision. In terms of the relationship between the size of the company and final decision-making, only the British/Irish data were significantly correlated ($r = -0.200$ at $\alpha = 0.05$). In general, as firm size increases management awareness as an influence and the need for cost considerations diminish (Table 4.31). Firms in size band 1-10 considered cost to be a relatively significant influence on the final decision. These firms may be referring to cost as a negative driver i.e. they are cost constrained. Management awareness was more important in firms employing fewer than 50 staff compared with larger firms.

Table 4.31. Influences on final decision regarding environmental issues by size, number and percentage (in brackets) of firms in each size group, UK/ROI, PQ

Size group	Regulation	Management awareness	Cost	Customer demands	Total no. of firms
1-10	7 (50.0)	3 (21.4)	4 (28.6)	–	14
11-25	6 (40.0)	5 (33.3)	3 (20.0)	1 (6.7)	15
26-50	7 (33.3)	8 (38.1)	3 (14.3)	3 (14.3)	21
51-100	14 (60.9)	4 (17.4)	2 (8.7)	3 (13.0)	23
101-250	12 (52.2)	5 (21.7)	2 (8.7)	4 (17.4)	23
251-500	5 (83.8)	–	1 (16.7)	–	6
Total	51 (50.0)	25 (24.5)	15 (14.7)	11 (10.8)	102

Table 4.32. Perception of the main impact of environmental legislation, number and percentage (in brackets) of firms, postal questionnaire

	Compliance only	Compliance plus	Excellence	All firms
UK/ROI				
Reduced competitiveness	17 (32.1)	10 (25.6)	4 (36.4)	31 (30.1)
Improvement for the good of society	21 (39.6)	13 (33.3)	4 (36.4)	38 (36.9)
Improved productivity	15 (28.3)	16 (41.0)	3 (27.3)	34 (33.0)
Total firms	53	39	11	103
Germany				
Reduced competitiveness	14 (56.0)	24 (49.0)	4 (30.8)	42 (48.3)
Improvement for the good of society	11 (44.0)	25 (51.0)	8 (61.5)	44 (50.6)
Improved productivity	–	–	1 (7.7)	1 (1.1)
Total firms	25	49	13	87
Italy				
Reduced competitiveness	4 (30.8)	22 (26.2)	–	26 (26.0)
Improvement for the good of society	7 (53.8)	58 (69.0)	3 (100.0)	68 (68.0)
Improved productivity	2 (15.4)	4 (4.8)	–	6 (6.0)
Total firms	13	84	3	100

Firms were asked to comment on the main impact of complying with environmental legislation (Table 4.32). In Germany and Italy there were a high proportion of negative responses concerning improved productivity. There was a general view in these countries that the main benefit would be to society as a whole. In the UK/ROI the answers were fairly evenly distributed between the three options. There were no significant differences in these responses between the environmental performance groups or between different sized firms.

Table 4.33. Environmental strategy, number and percentage (in brackets) of firms, FTF

	Compliance only	Compliance plus	Excellence	All firms
UK/ROI				
Regulation	13 (76.5)	2 (13.3)		15 (46.9)
Cost saving	1 (5.9)	4 (26.7)		5 (15.6)
Eco-practices	1 (5.9)	6 (40.0)		7 (21.9)
Eco-products	2 (11.8)	3 (20.0)		5 (15.6)
Total firms	17	15		32
Germany				
Regulation	3 (75.0)	4 (36.4)	2 (20.0)	9 (36.0)
Cost saving	1 (25.0)	5 (45.5)	2 (20.0)	8 (32.0)
Eco-practices	–	–	2 (20.0)	2 (8.0)
Eco-products	–	2 (18.2)	3 (30.0)	5 (20.0)
New markets for eco-products	–	–	1 (10.0)	1 (4.0)
Total firms	4	11	10	25

The responses from the face-to-face interviews regarding the strategy adopted by the company with respect to the environment are shown in Table 4.33. In the case of Italy the conclusions were the same as for the postal questionnaire results i.e. an exceptionally high proportion of firms (97%) endorsed regulation as their environmental strategy. In the UK/ROI and Germany the domination of regulation was still evident but not as markedly so. As with the findings from the postal questionnaire, there was clear evidence that when firms move beyond compliance they adopt environmental strategies that were proactive in terms of improving the market position of the firm by differentiating themselves from the competitors through eco-practices or eco-products. These differences between compliance and beyond compliance firms were statistically significant (at $\alpha = 0.01$ in the UK/ROI; $\alpha = 0.10$ in Germany).

4.9 Competitive Strategy

At the face-to-face interviews the managers were asked to describe the company's competitive strategy. On the basis of the responses the firms were characterised as belonging to one of the following: cost leadership with an emphasis on low cost, differentiation where the emphasis was on uniqueness, or niche with a focus on a particular segment of the market. Table 4.34 shows the resulting distribution. In the UK/ROI the firms interviewed adopted all three strategies in almost equal proportions. There was a difference between the two environmental performance groups in the type of strategy adopted. Cost leadership was the approach adopted by almost half of the compliance only firms compared with about one-quarter of the compliance plus firms. For firms following a differentiation strategy these proportions were reversed, and reduced for compliance only firms. In Italy the picture was similar to the distribution among the British/Irish compliance only firms. Since all the Italian firms except one were categorised as compliance only this position was not unexpected. In Germany the situation was somewhat different with no firms pursuing a low cost competitive strategy and a high proportion implementing a niche strategy irrespective of environmental performance.

Table 4.34. Competitive strategy adopted by environmental performance group, number and percentage (in brackets) of firms, face-to-face interviews

	Cost leadership	Differentiation	Niche	All firms
UK/ROI				
Compliance only	8 (47.1)	3 (17.6)	6 (35.3)	17 (100.0)
Compliance plus	4 (26.7)	7 (46.7)	4 (26.7)	15 (100.0)
All firms	12 (37.5)	10 (31.3)	10 (31.3)	32 (100.0)
Germany				
Compliance only	–	1 (11.1)	8 (88.9)	9 (100.0)
Compliance plus	–	5 (41.7)	7 (58.3)	12 (100.0)
Excellence	–	4 (36.4)	7 (63.6)	11 (100.0)
All firms	–	10 (31.2)	22 (68.8)	32
Italy				
All firms	13 (41.9)	6 (19.4)	12 (38.7)	31 (100.0)

Compliance plus firms were more likely to be pursuing a non-cost based competitive strategy. This may be because their ability to charge a price premium allows them the flexibility to adopt environmental initiatives or it may be that they consider that undertaking environmental initiatives reflects a form of differentiation/quality. Thus there was an expectation that firms following a cost leadership strategy would undertake fewer initiatives than those with a differentiation or niche strategy. However, the data did not support this assumption (Table 4.35).

Table 4.35. Competitive strategy and average number of eco-points, face-to-face interviews

	UK/ROI	Germany	Italy
Cost leadership	7.13	–	6.42
Differentiation	6.89	17.54	5.89
Niche	4.63	15.14	4.86

When comparing the relationship between the competitive strategy and the environmental strategy adopted by the firms there was an assumption that firms with a differentiation or niche competitive strategy would have had an environmental strategy that reflected this approach i.e. they would have implemented environmental strategies relating to environmental practices and products. The data in Table 4.36 shows that this was not the case and environmental strategies driven by regulation and cost savings were the predominant approaches regardless of the company's competitive strategy.

Table 4.36. Competitive and environmental strategy adopted, number of firms, FTF

	Cost leadership	Differentiation	Niche	Total
UK/ROI				
Regulation	7	4	4	15
Eco-efficiency	3	1	1	5
Eco-image	2	4	1	7
Eco-products	–	1	4	5
Eco-markets	–	–	–	–
Total firms	12	10	10	32
Germany				
Regulation		3	6	9
Eco-efficiency	–	3	5	8
Eco-image	–	1	1	2
Eco-products	–	2	3	5
Eco-markets	–	–	1	1
Total firms	–	9	16	25
Italy				
Regulation	11	6	11	28
Eco-efficiency	–	–	1	1
Eco-image	–	–	–	–
Eco-products	–	–	–	–
Eco-markets	–	–	–	–
Total firms	11	6	12	29

4.10 Summary

On average, German firms undertook more environmental initiatives than firms in the UK/ROI and in Italy. There was evidence that in the UK/ROI and Germany large firms implemented more environmental initiatives than small firms. There was some limited evidence in Germany and the UK/ROI to support the competitiveness hypothesis that a greater number of environmental initiatives were associated with greater competitiveness. There was no evidence to indicate that improved environmental performance led to a reduction in competitiveness.

In general, compliance only firms focused on waste separation and undertook virtually no management initiatives; compliance plus firms were involved in waste separation, chemical control and energy efficiency initiatives and the excellence firms were noted for their level of participation in management and eco-design activities. Cost was the main driver for the energy and packaging initiatives. Regulation and cost were the main drivers for waste management initiatives. Regulation and the market drove the more technically demanding initiatives. Initiatives that were related to design, materials change and environmental management systems were principally driven by the market. The surveys showed that there were economic benefits associated with environmental initiatives and that compliance only firms were able to achieve similar benefits as the better performing firms.

Many of the firms adopted a regulation strategy with respect to the environment. Compliance plus and excellence firms adopted a mix of strategies that included regulation, cost savings and proactive market strategies which impacted on both practices and processes. There was no evidence to suggest that the firms that pursued a non-cost based competitive strategy undertook more environmental initiatives.

5 Textile Finishing Industry

5.1 Introduction

The textile finishing industry (NACE code 17.3) includes activities such as bleaching, dyeing, finishing and printing and accordingly the firms sampled were primarily involved in wet manufacturing processes. The postal questionnaire (PQ) was distributed to 681 firms in Germany and 631 firms in the UK/ROI that were identified as belonging to this sector. Standard sources of information such as trade and telephone directories, business information publications and Internet databases were used to identify the population. The number of usable replies received was 73 for Germany and 78 for the UK/ROI. In Italy this short questionnaire was administered using the same approach as with the furniture industry. Starting with the larger firms in the Veneto region, firms were contacted by telephone until the target of 100 responses was achieved.

The choice of firms for the in-depth interviews (FTF) was made by careful consideration of the type of product manufactured, the size of the company and the extent to which, on the basis of information available, the firm could be categorised as compliance only or beyond compliance in terms of its environmental performance. The type of activities undertaken by these firms included dyeing, printing, bleaching and finishing of wool, cotton, flax and synthetic fabrics. In Italy, because of the absence of sufficient firms undertaking solely wet processes in the selected district, the Italian sample included 10 firms whose final product included knitted goods.

5.2 Firm Characteristics

The size distribution for the firms replying to the postal questionnaire and for firms interviewed face-to-face is shown in Tables 5.1 and 5.2. For the postal questionnaire the average employment for the UK/ROI was 92, for Germany 96 and for Italy, where the distribution was concentrated in the under 50 employees category, the comparable figure was 43. For the face-to-face sample, the average employment for the UK/ROI was 167, Germany 102 and Italy 83. In all countries there was no statistically significant difference in the average firm size between the PQ and FTF samples.

Table 5.1. Distribution of firms by size group, postal questionnaire

Size groups	UK/ROI		Germany		Italy	
	Number of firms	% of firms	Number of firms	% of firms	Number of firms	% of firms
1-10	12	15.4	12	16.9	19	19.6
11-25	13	16.7	16	22.5	36	36.0
26-50	17	21.8	9	12.7	25	25.0
51-100	17	21.8	10	14.1	10	10.0
101-250	11	14.1	18	25.4	8	8.0
251-500	6	7.7	6	8.5	2	2.0
500-599*	2	2.6	–	–	–	–
Total	78	100.0	71	100.0	100	100.0

* Neither of these firms were considered to be outliers.

Table 5.2. Distribution of firms by size group, face-to-face interviews

Size groups	UK/ROI		Germany		Italy	
	Number of firms	% of firms	Number of firms	% of firms	Number of firms	% of firms
1-10	1	3.0	3	10.0	5	15.2
11-25	3	9.1	4	13.3	8	24.2
26-50	5	15.2	6	20.0	7	21.2
51-100	6	18.2	5	16.7	3	9.1
101-250	11	33.3	9	30.0	8	24.2
251-500	7	21.2	3	10.0	2	6.1
Total	33	100.0	30	100.0	33	100.0

For each country, Table 5.3 shows the summary data for the responses to the postal questionnaire regarding the environmental initiatives undertaken. Also shown is the number of firms and the average number of initiatives carried out in each of the environmental performance groups (see Sect. 2.2 for details of these initiatives and performance groups).

On the basis of the overall count, Germany undertook considerably more initiatives than either of the other countries. There was strong evidence to support a statistically significant difference in the average number of initiatives carried out between the countries (at $\alpha = 0.01$). When these data were reclassified into the three environmental performance groups, there was no evidence to suggest any difference between the countries in the average number of initiatives carried out for either the compliance only or excellence categories. Hence compliance only and excellence firms had the same environmental performance in the UK/ROI as in the other two countries. Conversely for the compliance plus firms there was sufficient evidence to reject the null hypothesis of equality (at $\alpha = 0.01$) in the average number of initiatives undertaken between the countries, showing that compliance plus firms in Germany had better environmental performance than compliance plus firms in the other countries.

Table 5.3. Sample size, number of initiatives undertaken and environmental performance groups, postal questionnaire

	UK/ROI		Germany		Italy	
Sample size	78		73		100	
Total number of initiatives	388		454		211	
Average number of initiatives	4.97		6.22		2.11	
Environmental Performance Groups						
	No. of firms	Av. No. of initiatives	No. of firms	Av. No. of initiatives	No. of firms	Av. No. of initiatives
Compliance only	15	1.53	9	1.22	67	1.46
Compliance plus	29	3.86	18	4.50	32	3.34
Excellence	34	7.44	46	7.87	1	6.00

The face-to-face interviews with firms (FTF) enabled a more complete analysis of the extent and quality of environmental initiatives undertaken. Firms were asked detailed questions about eight initiatives that were particular to the textile finishing industry and for each SME an eco-point score was calculated (see Sect 2.2 for details).

Table 5.4. Sample size, eco-points and environmental performance groups, FTF interviews

	UK/ROI		Germany		Italy	
Sample size	33		30		33	
Total number of eco-points	282		521		147	
Average number of eco-points	8.53		17.37		4.45	
Environmental Performance Groups						
	No. of firms	Av. No. of eco-points	No. of firms	Av. No. of eco-points	No. of firms	Av. No. of eco-points
Compliance only	16	5.25	6	10.00	6	1.50
Compliance plus	17	11.36	11	16.06	20	4.10
Excellence	–	–	13	21.90	7	8.00

Table 5.4 shows that German firms undertook the highest average number of initiatives followed by firms in the UK/ROI. The Italian data was not strictly comparable as it contained firms that were involved in activities additional to wet processes i.e. final clothing products. There was strong evidence of a statistically significant difference (at $\alpha = 0.01$) in the average number of eco-points between all the countries. This difference was maintained when these data were disaggregated into environmental performance groups.

A priori, there was the expectation that in terms of the average number of initiatives undertaken, which was a proxy for environmental performance, Germany would rank first, followed by the UK/ROI and finally Italy. The postal responses and face-to-face interviews confirmed that this was indeed the case.

The respondents to the FTF interview were asked to characterise their current profitability and expectations of growth in the next three years (Table 5.5). In terms of current profitability the German and the British/Irish companies were reasonably buoyant with over 60% of firms currently operating in profit. The situation in Italy was somewhat less positive with only 40% of firms in profit. In the UK/ROI firms maintained this pattern irrespective of environmental performance whereas in Germany a much higher proportion of the excellence firms stated that they were operating in profit (92%). In Italy almost all the compliance only firms indicated that they were operating at a loss or just breaking even. The expectation of growth in sales over the next three years was, on the whole, positive with a reasonable proportion of firms signifying increased sales. This was encouraging, as the textile finishing industry has been facing difficulties with competition from non-EU countries. In all the countries there was no difference in expected growth between the environmental performance categories.

Table 5.5. Economic performance of SMEs, face-to-face interviews

	UK/ROI		Germany		Italy	
	Number of firms	% of firms	Number of firms	% of firms	Number of firms	% of firms
Net profit						
Loss	3	9.1	7	23.3	2	6.4
Break-even	9	27.3	3	10.0	16	52.6
Profit 1-10%	19	57.6	15	50.0	12	38.7
Profit >10%	2	6.0	5	16.7	1	3.2
Total	33	100.0	30	100.0	31	100.0
Expected growth						
Reduced	4	13.3	9	30.0	3	9.4
Stable	8	26.6	8	26.6	11	34.3
Increased	18	60.0	13	43.4	18	56.3
Total	30	100.0	30	100.0	31	100.0

At interview, managers were asked to identify the factors that they considered to be vital for competitiveness in their business. Price and quality were judged to be the two most important factors. Factors such as service, design and labour quality were rated as being of secondary importance while capital availability; marketing and the environment and consultancy services were not thought to have an influence on firm competitiveness. This rating was consistent across the countries.

5.3 Size Effect

Table 5.6 shows the relationship between firm size and environmental performance, measured as the average number of initiatives per firm for the data from the postal questionnaire and as the average number of eco-points per firm for data from the face-to-face interviews. In terms of the PQ data, there was a statistically significant relationship between size and the number of initiatives undertaken in all the countries. The correlation coefficients were; for the UK/ROI 0.361, significant at $\alpha = 0.01$; for Germany 0.251 at $\alpha = 0.05$ and for Italy 0.296 at $\alpha = 0.01$. This was reinforced by testing for differences in the employment means between the environmental performance groups, which were significant in all the countries.

The results from the FTF interviews (Table 5.6) showed a positive linear relationship between the size of the firm and environmental performance ($r = 0.473$, at $\alpha = 0.01$) in the UK/ROI. However, as expected there was no significant difference in average employment between the compliance only and compliance plus groups because these data have been matched hence the effect of size has been controlled for. Neither the German nor Italian datasets showed a relationship between the size of the firm and environmental performance (as measured by the number of eco-points).

Table 5.6. Average number of initiatives (PQ) and eco-points (FTF) by size group

Size groups	UK/ROI		Germany		Italy	
	PQ	FTF	PQ	FTF	PQ	FTF
1-10	4.00	2.00	3.42	11.00	2.10	3.00
11-25	4.69	6.33	5.94	15.00	1.80	4.75
26-50	3.94	6.13	6.67	16.67	2.00	4.86
51-100	4.82	6.89	6.90	17.20	2.60	4.67
101-250	6.73	9.82	7.56	21.15	3.00	5.00
251-500	6.67	11.52	5.83	17.33	3.00	3.00
500-599	8.00	–	–	–	–	–
Total	4.97	8.53	6.22	17.37	2.11	4.45

5.4 Competitive Status

The indicators of competitiveness collected in the postal questionnaire were firm profitability, value added per employee (except for Italy), employment in R&D, employment growth in the previous five-year period and destination of sales.

Firm profitability and value-added per employee were significantly correlated with one another in the UK/ROI and Germany ($r = 0.598$ and $r = 0.436$ respectively with both coefficients significant at $\alpha = 0.01$). In Italy firm profitability was positively correlated with the number of environmental initiatives undertaken ($r = 0.199$ at $\alpha = 0.05$) but this relationship was not sustained when the effect of firm size was controlled for. Neither profitability nor value added were correlated with the number of initiatives undertaken in the UK/ROI and Germany (Table 5.7). In addition, these variables showed no significant difference between the environmental performance groups (Tables 5.8 and 5.9). Hence, good environmental performance did not depend on the firms' economic performance as measured by these two variables.

Table 5.7. Firm profitability and value added per employee, average number of initiatives, postal questionnaire

Compared to industry average	Firm Profitability	Value added per employee
UK/ROI		
Below average	4.95	5.00
Average	4.39	4.89
Above average	6.83	4.89
Germany		
Below average	6.71	6.53
Average	5.81	6.08
Above average	6.50	6.67
Italy		
Below average	1.57	No data
Average	2.10	No data
Above average	2.67	No data

Table 5.8. Firm profitability, number and percentage (in brackets) of firms, PQ

Compared to industry average	Compliance (0-2 inits)	Compliance + (3-5 inits)	Excellence (6-11 inits)	All firms
UK/ROI				
Below average	3 (20.0)	9 (34.6)	9 (27.3)	21 (28.4)
Average	11 (73.3)	15 (57.7)	15 (45.5)	41 (55.4)
Above average	1 (6.7)	2 (7.7)	9 (27.3)	12 (16.2)
Total firms	15	26	33	74
Germany				
Below average	2 (22.2)	5 (29.4)	17 (37.0)	24 (33.3)
Average	6 (66.7)	9 (52.9)	21 (45.7)	36 (50.0)
Above average	1 (11.1)	3 (17.7)	8 (17.4)	12 (16.7)
Total firms	9	17	46	72
Italy				
Below average	5 (7.6)	2 (6.5)	–	7 (7.1)
Average	56 (84.4)	25 (80.6)	1 (100.0)	82 (83.7)
Above average	5 (7.6)	4 (12.9)	–	9 (9.2)
Total firms	66	31	1	98

Table 5.9. Value added per employee, number and percentage (in brackets) of firms, PQ

Compared to industry average	Compliance (0-2 inits)	Compliance + (3-5 inits)	Excellence (6-11 inits)	All firms
UK/ROI				
Below average	1 (6.7)	5 (19.2)	5 (15.6)	11 (15.1)
Average	10 (66.7)	14 (53.8)	20 (62.5)	44 (60.3)
Above average	4 (26.7)	7 (26.9)	7 (21.9)	18 (24.7)
Total firms	15	26	32	73
Germany				
Below average	3 (33.3)	1 (5.8)	10 (22.7)	14 (20.0)
Average	6 (66.7)	13 (76.5)	31 (70.5)	50 (71.4)
Above average	–	3 (17.7)	3 (6.8)	6 (8.6)
Total firms	9	17	44	70

No data collected for Italy.

The postal questionnaire collected data on the extent of R&D activity in the firm. The measures used to test the relationship between R&D and environmental performance were the number of full-time equivalent employees[18] engaged in R&D and the number of initiatives undertaken. There was significant positive correlation between employment in R&D and environmental performance (Table 5.10), although fairly weak in the case of the UK/ROI.

Table 5.10. R&D staff (full-time equivalents), postal questionnaire

Number of R&D staff (FTEs)	Average number of initiatives		
	UK/ROI	Germany	Italy
0	3.60	4.76	1.74
0.5-2.5	5.04	6.67	2.46
3.0-5.5	4.07	7.62	3.43
6.0-10.5	8.00	6.00	3.00
11.0-20.5	4.00	6.50	–
21+	8.50	11.00	–
Correlation coefficient (R&D FTEs and number of initiatives)	0.215*	0.244**	0.414***

*** significant at 0.01; ** significant at 0.05; * significant at 0.1.

It was shown earlier that there was a positive linear relationship between the number of initiatives and firm size and an association with firm size was also evident when considering the number of staff engaged in R&D activities (Table 5.11). Because of the strong correlations between firm size and the numbers employed in R&D functions, the data was tested for evidence of correlation between R&D and environmental performance while controlling for the effect of firm size. The re-

[18] A full-time equivalent measure was constructed by counting an employee working full-time in R&D as one and counting those working part-time in R&D activities as 0.5.

sults showed that in all the countries there was no significant relationship. However this does not mean the R&D had no impact on the ability of the firm to introduce environmental initiatives.

Table 5.11. Employment in R&D by firm size, postal questionnaire

Size groups	Average number of FTEs in R&D		
	UK/ROI	Germany	Italy
1-10	1.21	0.13	0.76
11-25	1.04	0.66	0.49
26-50	1.50	0.89	1.28
51-100	2.50	2.75	2.05
101-250	5.00	3.11	1.75
251-500	8.58	9.00	4.75
500-599	17.00	–	–
All firms	3.03	2.17	1.08
Correlation coefficient (employment and R&D FTEs)	0.649***	0.589***	0.316***

*** significant at 0.01.

Labour growth, as measured by the change in employment in the firm over the previous five-year period, was not found to have a statistically significant relationship with environmental performance in any of the countries.

The final measure of competitiveness, from the postal questionnaire data, was a ranking based on the destination sales (local, national and international). For each country, Table 5.12 gives details of the criteria used, the firm distribution and the average number of initiatives undertaken. As there was no statistically significant difference between countries in the average level of exports the decision to apply the same criteria to all countries was taken.

The Italian data showed no correlation between environmental performance and destination of sales. In the UK/ROI there was a significant positive relationship ($r = 0.261$ at $\alpha = 0.05$) between environmental performance and destination of sales. However, the latter was also correlated with firm size ($r = 0.395$ at $\alpha = 0.01$) and when firm size was taken into account, the relationship between environmental performance and destination of sales was no longer statistically significant. In Germany this competitiveness measure and environmental performance were positively correlated ($r = 0.381$ at $\alpha = 0.01$) and there was also a relationship between the ranking and firm size ($r = 0.532$ at $\alpha = 0.01$). However, in contrast to the British/Irish case, this positive connection between destination of sales and environmental performance remained valid when firm size was controlled ($r = 0.388$ at $\alpha = 0.01$). Hence larger SMEs in the UK/ROI, which sell nationally and internationally, were better environmental performers. In Germany, the same relationship was valid but was applicable to all employee size groups. The variability in the relationship between exports and environmental performance in the countries can also be seen from the data in Table 5.13 where in Germany the firms that exported were to be found primarily in the excellence environmental performance category.

Table 5.12. Sales destination ranking, postal questionnaire

Rank	Rule	UK/ROI		Germany		Italy	
		No. of firms	No. of inits.	No. of firms	No. of inits.	No. of firms	No. of inits.
1	No exports, national sales < 50%	12	3.67	16	4.19	30	1.93
2	No exports, national sales >= 50%	10	4.60	5	3.40	13	2.46
3	Exports < 10%	14	5.14	13	7.62	8	2.00
4	Exports 10 - 44%	19	4.58	28	6.61	21	2.29
5	Exports >= 45%	20	6.00	9	7.56	28	2.04

Table 5.13. Competitiveness ranking and environmental performance, postal questionnaire

	All firms	Compliance only	Compliance plus	Excellence
	No. (% of firms)	% of firms	% of firms	% of firms
UK/ROI				
No exports, national sales < 50%	12 (16.0)	25.0	58.3	16.7
No exports, national sales >= 50%	10 (13.3)	20.0	40.0	40.0
Exports < 10%	14 (18.7)	14.3	35.7	50.0
Exports 10% - 44%	19 (25.3)	31.6	31.6	36.8
Exports >= 45%	20 (26.7)	10.0	25.0	65.0
Germany				
No exports, national sales < 50%	16 (22.5)	31.3	25.0	43.8
No exports, national sales >= 50%	5 (7.0)	40.0	40.0	20.0
Exports < 10%	13 (18.3)	–	15.4	84.6
Exports 10% - 44%	28 (39.4)	3.6	32.1	64.3
Exports >= 45%	9 (12.7)	11.1	11.1	77.8
Italy				
No exports, national sales < 50%	30 (30.0)	70.0	30.0	–
No exports, national sales >= 50%	13 (13.0)	53.8	46.2	–
Exports < 10%	8 (8.0)	75.0	25.0	–
Exports 10% - 44%	21 (21.0)	66.7	28.6	4.8
Exports >= 45%	28 (28.0)	67.9	32.1	–

On the basis of the postal questionnaire data alone, there was little evidence to support the hypothesis that more competitive firms undertake more environmental initiatives. Only in Germany was there some evidence that firms that serve an export market did in fact carry out more environmental initiatives.

In addition to data on the destinations of sales, employment growth and participation in R&D, the face-to-face interviews collected data on the following explanatory variables – productivity per head, age of machinery, and staff qualifications.

Table 5.14. Correlation between measures of competitiveness and environmental performance, face-to-face interviews

	Expected sign	Correlation coefficient	Level of significance
UK/ROI			
Exports (%)	+	0.378	0.05
Age of Machinery	-	0.020	ns
Productivity (turnover/capita)	+	-0.048	ns
R&D (% of total employment)	+	-0.310	ns
Growth (employment, 1993-1998)	+	-0.029	ns
Qualifications (% of total employment)	+	0.067	ns
Germany			
Exports (%)	+	-0.054	ns
Age of Machinery	-	-0.472	0.01
Productivity (turnover/capita)	+	-0.204	ns
R&D (% of total employment)	+	0.288	ns
Growth (employment, 1993-1998)	+	0.155	ns
Qualifications (% of total employment)	+	0.065	ns
Italy			
Exports (%)	+	0.055	ns
Age of Machinery	-	-0.191	ns
Productivity (turnover/capita)*	+	n.a.	
R&D (% of total employment)	+	0.018	ns
Growth (employment, 1993-1998)	+	0.023	ns
Qualifications (% of total employment)	+	-0.243	ns

* Data on total productivity per firm and hence productivity per head were not available as variables in the Italian dataset.

On the whole, the correlation analysis of the relationship between these measures of competitiveness and environmental performance for the FTF interviews (Table 5.14) showed that there were no significant relationships. The exceptions were exports in the UK/ROI and the age of machinery in Germany. In the UK/ROI, as with the PQ data, the relationship between exports and environmental performance was complicated by the effect of firm size. Firm size was significantly correlated with exports ($r = 0.383$ at $\alpha = 0.05$) and when the effect of size was taken into account exports and environmental performance did not show a significant correlation. In Germany the age of machinery had a direct bearing on the level of environmental performance and this relationship remained valid when firm size was taken into account ($r = -0.464$ at $\alpha = 0.05$). Tables 5.15 to 5.17 show these measures of competitiveness by environmental performance group for each country.

Table 5.15. Average competitiveness measures by environmental performance groups, UK/ROI, face-to-face interviews

	Compliance only	Compliance plus	All firms
Av. Turnover (£ m)	10.460	15.150	12.810
Av. Productivity per head (£)	77,260	85,740	81,500
Av. Level of exports (%)	39.7	47.6	43.7
Av. Employment per firm	147	187	167
Av. Employment growth (%)	-11.2	5.28	-2.67
Av. Development staff per firm	1.9	3.2	2.6
Av. Qualified staff (% of all staff)	16.5	16.0	16.2
Av. Wages (£ per hour)	5.77	6.20	5.95
Av. Age of machinery (years)	11.57	12.44	11.98
Sample size	16	17	33
Average number of eco-points	5.25	11.63	8.53

Table 5.16. Average competitiveness measures by environmental performance groups, Germany, face-to-face interviews

	Compliance only	Compliance plus	Excellence	All firms
Av. Turnover (€, million)	6.250	9.800	8.760	8.720
Av. Productivity per capita (€)	116,000	72,000	69,700	78,700
Av. Level of exports (%)	16.5	15.2	18.9	17.0
Av. Employment per firm	58	110	115	102
Av. Employment growth (%)	-14.9	-10.0	3.2	-4.6
Av. Development staff per firm	0.0	1.8	3.5	2.2
Av. Qualified Staff (% of all staff)	51.8	46.7	57.6	52.1
Av. Wage (€ per hour)	11.2	9.6	9.8	10.0
Av. Age of machinery (years)	11.8	11.2	8.3	10.1
Sample size	6	11	13	30
Average number of eco-points	10.00	16.06	21.90	17.37

Table 5.17. Average competitiveness measures by environmental performance groups, Italy, face-to-face interviews

	Compliance only	Compliance plus	Excellence	All firms
Av. Turnover (€, million)*	11.227	13.189	20.041	14.820
Av. Productivity per capita (€)	n.a.	n.a.	n.a.	n.a.
Av. Level of exports (%)	11.3	21.9	23.6	20.3
Av. Employment per firm	103	82	72	83
Av. Employment growth (%)	2.4	106.9	75.3	84.6
Av. Development staff per firm	0.33	3.90	2.71	3.0
Av. Qualified staff (% of all staff)	75.3	58.4	49.0	59.5
Av. Wage (€ per hour)*	30.0	30.0	30.0	30.0
Av. Age of machinery (years)	7.48	4.64	4.54	5.14
Sample size	6	20	7	33
Average number of eco-points	1.50	4.10	8.00	4.45

* Estimates based on interviews with sector experts as opposed to firm managers.

There was no significant difference in these average measures of competitiveness between the environmental performance groups in either the British/Irish or the Italian sample data. In the German case there was a significant difference between performance groups for the age of machinery and R&D variables.

In summary, there was limited evidence to support the hypothesis that firms with above average economic performance will also be superior in terms of environmental performance. In the UK/ROI, the size of the firm had a key role in determining economic and environmental performance. Even when the firms were matched in terms of size there was no discernable difference between compliance and compliance plus firms. In Italy there was no observable relationship between economic and environmental performance. However, in Germany, up-to-date equipment, employment in R&D and servicing an export market had a noticeable influence on environmental performance.

5.5 Environmental Initiatives

The adoption of the individual environmental initiatives by firms in the PQ sample in the countries studied was variable (Fig. 5.1). In all the countries, the type of initiative most frequently undertaken was related to waste management, particularly waste separation. The reuse of waste was equally important in the UK/ROI and Germany but considerably less so in Italy. In an industry that makes extensive use of water and chemicals (such as dyes, bleaches) in the production cycle, initiatives aimed at increasing efficiency in energy, water and chemical use were implemented by at least half of all the British/Irish and German firms. The more sophisticated method of reducing energy costs through integrated heat and water systems had a much smaller uptake. Market focused initiatives, such as product labelling

(Eco-Tex), were important in Germany (with 55% of firms partaking) and to a lesser extent in Italy (20%) and the UK/ROI (12%). Co-operation along the supply chain as a measure to increase sales was fairly common in both the UK/ROI and Germany. Management initiatives that would lead to improved environmental accounting and procedures were adopted by just over half of the UK/ROI companies, one-quarter of the German firms and were virtually absent in Italy. Italian firms did not participate in initiatives that were designed to improve efficiency through either reducing energy costs or limiting the use of chemicals in the manufacturing processes, a reflection of the mixed activities surveyed.

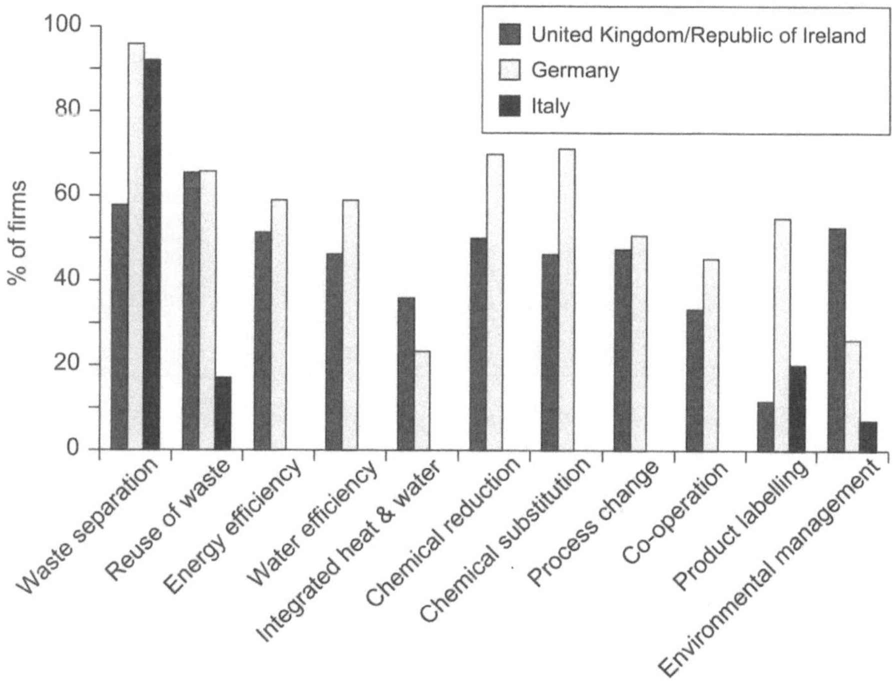

Fig. 5.1. Proportion of firms undertaking each initiative, all countries, postal questionnaire

Not only was there variation in the type of initiative carried out at the national level, there was also considerable dissimilarity in the type of initiative undertaken by the three environmental performance groups (Figs. 5.2 and 5.3). In all the countries, compliance only firms focused on waste management; undertook few efficiency initiatives and almost none of the firms carried out any market or management related activities. In Germany and the UK/ROI, compliance plus firms indicated their involvement in almost all the initiatives but to differing degrees. For example, almost 35% of the British/Irish compliance plus firms participated in management initiatives compared to only 18% of Italian firms and 5% of German companies. Excellence firms were distinguished from compliance only and compliance plus firms in that, in addition to waste management and efficiency measures, they also undertook activities that were market driven or management focused. Differences across the environmental performance groups were statistically significant for all eleven listed initiatives in Germany and, with the exception of product labelling, in the UK/ROI also. In Italy, which undertook only four[19] of the listed initiatives, there was no difference across performance groups for waste separation but there was for the other three initiatives with a significant difference between excellence and other firms.

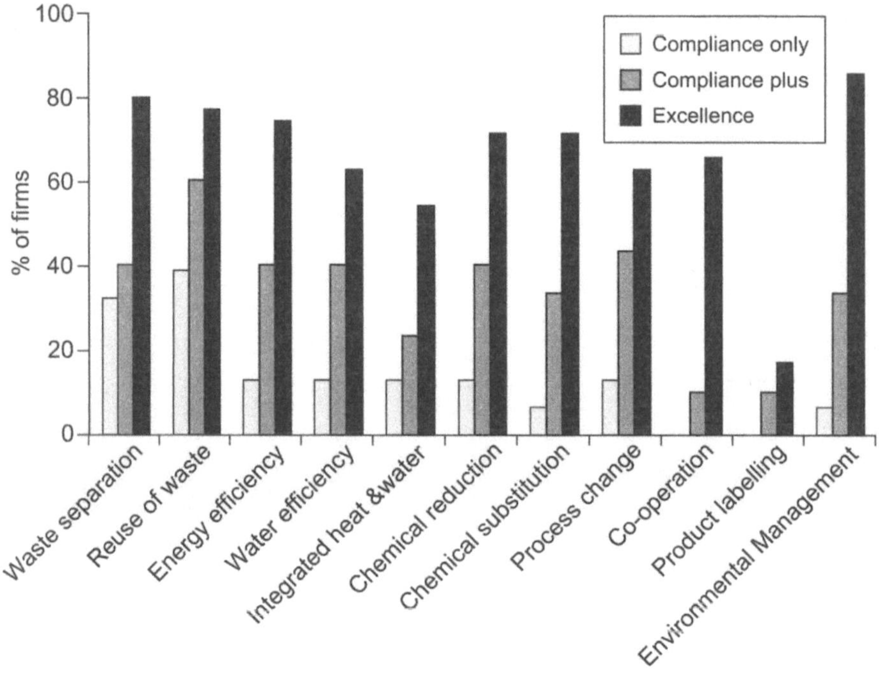

Fig. 5.2. Proportion of firms undertaking each initiative, UK/ROI, postal questionnaire

In general, as firms moved up the scale of environmental performance they graduated from solely "housekeeping" type initiatives, such as waste management, to activities that required technical expertise and ultimately reached a level where organisational and market focused initiatives were considered to be on a par with the less complex activities. As was shown in Sect. 5.3, there was a relationship between the size of the firm and the number of initiatives undertaken. On the whole, the smaller firms favoured waste management initiatives and as size increased proportionately more firms undertook efficiency, market and management initiatives.

[19] Reuse of waste, waste separation, product labeling and environmental management.

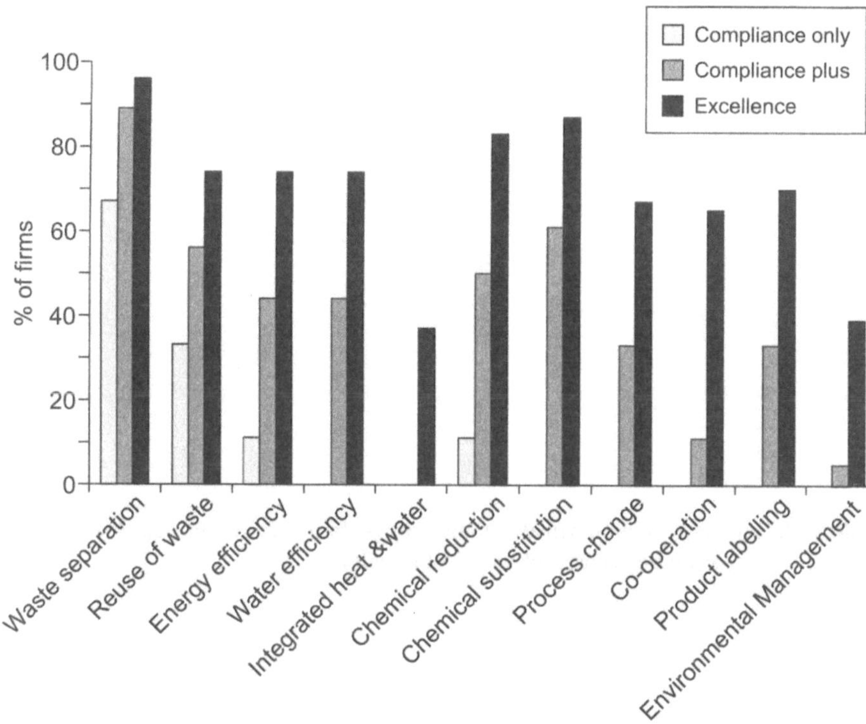

Fig. 5.3. Proportion of firms undertaking each initiative, Germany, postal questionnaire

In the UK/ROI, the results from the face-to-face interviews (Table 5.18) followed the pattern observed in the PQ sample. Compliance plus firms undertook more initiatives in the areas of energy efficiency, water protection, packaging, environmental management and communication than compliance only firms. There was a statistically significant difference (at $\alpha = 0.05$) between the two environmental performance groups for these five initiatives. Chemical substitution and reduction showed a significant difference at $\alpha = 0.10$ whereas waste separation and air protection did not register any difference. A high proportion of firms, irrespective of performance group, undertook waste separation whilst few firms carried out air protection initiatives.

Table 5.18. UK/ROI: Environmental initiatives by performance groups, FTF interviews

Initiative	Compliance only		Compliance plus		All firms	
	No. of firms	%	No. of firms	%	No. of firms	%
Energy efficiency	11	68.8	17	100.0	28	84.8
Environmental management	9	56.2	17	100.0	26	78.7
Waste separation	13	81.2	12	70.6	25	75.6
Water use and protection	8	50.0	14	82.4	22	66.7
Chemical substitution and/or reduction	8	50.0	13	76.4	21	63.6
Packaging	6	37.5	14	82.4	20	60.6
Communication	–		5	35.2	5	15.1
Air protection	1	6.2	3	17.6	4	12.1
Sample size	16		17		33	

Table 5.19. Germany: Environmental initiatives by performance groups, face-to-face interviews (figures in brackets are % of sample)

Initiative	Compliance only	Compliance plus	Excellence	All firms
Energy efficiency	3 (50.0)	9 (81.8)	13 (100)	25 (83.3)
Environmental management*	–	2 (18.2)	8 (61.5)	10 (33.3)
Waste separation	5 (83.3)	11 (100)	13 (100)	29 (96.7)
Water use and protection	3 (50.0)	11 (100)	11 (84.6)	25 (83.3)
Chemical substitution and/or reduction	3 (50.0)	10 (90.9)	13 (100)	26 (86.7)
Packaging	5 (83.3)	11 (100)	12 (92.3)	28 (93.3)
Communication	4 (66.7)	10 (90.9)	12 (92.3)	26 (86.7)
Air protection	1 (16.7)	4 (36.4)	6 (46.2)	11 (36.7)
Sample size	6	11	13	30

* Only companies that had implemented a management system such as ISO9000 or 14001 were included.

In broad terms, the distribution of firms in Germany that took part in the interviews followed a fairly similar pattern to the PQ in terms of specified initiatives (Table 5.19). There were some discrepancies but this may be a reflection of the different survey methods, whereby, at interview the firms could explain in some detail the type of initiatives that were carried out. As in the UK/ROI, the excellence performance firms had a high level of participation in all the initiatives whereas, with the exception of waste separation and packaging, the uptake of initiatives amongst compliance only firms was at a fairly low level. There was a statistically significant difference (at $\alpha = 0.05$) between the performance groups in terms of management, energy, chemical and water initiatives. Excellence firms were distinguishable from the other groups in that a high proportion had environmental systems actually in operation.

The Italian firms sampled exhibited a similar pattern to the above (Table 5.20).

Table 5.20. Italy: Environmental initiatives by performance groups, face-to-face interviews (figures in brackets are % of sample)

Initiative	Compliance only	Compliance plus	Excellence	All firms
Energy efficiency	5 (83.3)	19 (95.0)	6 (85.7)	30 (90.9)
Environmental management	1 (16.7)	5 (25.0)	5 (71.4)	11 (33.3)
Waste separation	6 (100.0)	16 (80.0)	5 (71.4)	27 (81.8)
Eco materials	–	12 (60.0)	6 (85.7)	18 (54.5)
Packaging	–	8 (40.0)	3 (42.9)	11 (33.3)
Environmental technologies	1 (16.7)	17 (85.0)	6 (85.7)	24 (72.7)
Sample size	6	20	7	33

Data on the individual environmental initiatives were collected during the face-to-face interviews and the remainder of this section presents detailed findings. There are constraints with these data in that not all firms provided the information requested and there was little comparable data for Italy.

5.5.1 Waste Reduction

A substantial proportion of the companies in the sample carried out waste management initiatives. Table 5.21 shows, for the UK/ROI and Germany, the various waste streams and the number of firms participating in each activity. German firms were much more active in separating waste and a high proportion of the non-chemical waste went to recycling rather than disposal. In the UK/ROI waste management was concentrated on paper, cardboard and waste textiles.

Data relating to the level of financial investment involved to facilitate waste separation were available only for some of the German sample firms. Amounts invested ranged from €3,000 to €10,000 mostly for the purchase of compactors. The motivation for separating waste was predominately to achieve cost savings by reducing landfill costs, particularly in the UK/ROI where such charges have been steadily increasing. As well as reducing costs some firms had generated a small but growing amount of revenue by selling waste textiles on to recyclers.

Table 5.21. Waste management, UK/ROI and Germany, number of firms, FTF interviews

Waste stream	UK/ROI	Germany
Steel waste	9	23
Paper and cardboard	21	27
PE-film	6	25
Waste textiles for recycling	18	23
Wood	8	13
Waste chemicals (special waste)	6	12
Filter cake or sludge	2	13
Oily materials	8	11
Household type waste	–	27
Others	–	4
Sample size	33	30

5.5.2 Energy Initiatives

Table 5.22 shows the uptake of the various energy efficient initiatives in the UK/ROI and Germany. Comprehensive measures to minimise energy cost were mostly undertaken by the better environmental performing firms. This was especially the case for the sophisticated systems of energy efficiency such as integrated heat and water management. Such initiatives required a substantial capital investment ranging from £10,000 to £70,000 in the UK/ROI and from €3,000 to €750,000 in Germany. The payback period for such investment was, on average, between 2 and 4 years. In general, this type of technology was installed in the mid to late 1990s so in many cases firms were already experiencing benefits in terms of cost savings which were, on the whole, rated as substantial.

Table 5.22. Energy initiatives, UK/ROI and Germany, number of firms, FTF interviews

Initiative	UK/ROI	Germany
Efficient heat & power production e.g. combined heat and power plant	4	3
Efficient energy distribution	7	5
Efficient energy use or re-use e.g. integrated heat and water management	11	23
Energy efficient lighting	11	6
Other	6	2
Sample size	33	30

Energy efficient lighting systems were less expensive to introduce (around £5,000) and had a reasonably high level of implementation in the British/Irish firms. For this technology, compliance only firms were as likely to install these systems as their better performing counterparts. The level of cost saving achieved was rated as medium to high by the firms that implemented such technology.

Other methods of reducing energy costs ranged from, at the expensive end of the scale, efficient heat management with installation costs in Germany of between €0.5 million and €2.5 million, through to installing managed start-up devices on machinery, switching from oil to gas as the primary fuel (especially in the British/Irish firms) and educating workers in the need to conserve energy.

The main motivation for introducing these initiatives was to achieve cost savings and firms that had invested in such initiatives did acknowledge a reduction in costs and hence increased profitability.

5.5.3 Packaging

In the textile finishing firms interviewed, packaging initiatives were considered as being of no real importance to overall environmental performance. In Germany textiles were rolled and wrapped in polyethylene for dispatch, whereas in the UK/ROI the use of cardboard rolls was the main method of packaging for dispatch (Table 5.23). British/Irish firms acknowledged that in this sector it was difficult to reduce packaging per unit of output and, in general, it was the compliance plus firms that addressed these issues.

Table 5.23. Packaging initiatives, UK/ROI and Germany, number of firms, FTF interviews

Initiative	UK/ROI	Germany
Cardboard rolls used several times	6	6
Cardboard rolls used once	5	6
PE-film	4	26
Paper	–	2
Other	4	9
Sample size	33	30

5.5.4 Avoidance of Toxic Substances

Table 5.24 shows the breakdown of the various initiatives employed by textile finishing firms in order to limit chemical usage. Reduced chemical usage was achieved through either substituting the questionable chemicals with alternatives or by applying changes to the manufacturing process. In Germany about two-thirds of the firms tried to reduce the amounts of questionable substances or to phase them out entirely. The equivalent figure in the UK/ROI was about 45%.

The move towards natural or ecological raw materials has been fairly recent and the following account, from the German sample, of introducing these natural materials into the production process highlights some of the difficulties surrounding such substitution. "Five years ago company X ran a project testing plant colours for industrial purposes. All in all they made four applications in their own laboratory. The results were rather disillusioning; the technique proved impractical (substantial residue left on the fabric, very smelly, limited colour-stability) and thus inappropriate for industrial use".

Table 5.24. Toxic reduction and material change in inputs, UK/ROI and Germany, number of firms, FTF interviews

Initiative	UK/ROI	Germany
Avoid unnecessary or questionable finishing (e.g mothproofing of carpets)	5	15
Substitution of sub-optimal or questionable processes (e.g. continuous dyeing is less pollutant than discontinuous)	3	21
Substitution of single substances (e.g. AZO-colours or heavy metal colours in dyeing or penta chlorophenol in rot-proofing).	14	23
Use-reduction of single substances (e.g. low liquor ratios means less pollution)	6	20
Other	2	2
Natural/Ecological materials		
Plant colours	–	2
Mineral colours	–	1
Others	1	5
Sample size	33	30

There was a distinct difference in the uptake of these initiatives by environmental performance group with the compliance only firms undertaking substantially fewer. The dominant driver for undertaking these initiatives was regulation in the UK/ROI and market pressures in Germany. In general, controlling chemical inputs was costly and the firms did not on the whole achieve any direct cost benefit but were able to record an increase in market position and image.

5.5.5 Air Protection

Air protection initiatives were motivated by mandatory legislation, environmental and health and safety, and were adopted by about one-third of German firms but by a much smaller proportion of British/Irish firms (12%). All the methods used to limit air emissions were end-of-pipe technologies and Table 5.25 shows that filtration systems were the most frequent. Data on investment cost was available only for Germany and showed that filtration and absorption had the lowest cost at around €50,000 with vapour incinerators the most expensive at around €250,000. The latter however were very effective in reducing emissions by 95-99% without a corresponding increase in running costs.

Table 5.25. Air protection, UK/ROI and Germany, number of firms, FTF interviews

Initiative	UK/ROI	Germany
Adsorption	1	1
Incineration	–	1
Filtration	4	5
Electrostatic precipitation	–	–
Wet scrubbing	1	3
Other	–	5
Sample size	33	30

5.5.6 Water Use and Protection

Since many processes in this industry involve water it was not surprising that a high proportion of firms had adopted some form of water treatment (67% of the firms in the UK/ROI and 84% in Germany). Water treatment plants were most likely to be found in compliance plus and excellence firms. Much of the technology used dated from the late 1990s and the associated investment costs were quite substantial. Table 5.26 shows that water treatment plants and water saving technology were equally prevalent in the British/Irish firms. The majority of the improvements were made for regulatory reasons to reduce discharge levels although some firms cited cost as the main driver. On the whole the impact of this technology on the British/Irish firms was slightly reduced costs and a small increase in competitiveness.

Table 5.26. Water use and protection, UK/ROI and Germany, number of firms, FTF interviews

Initiative	UK/ROI	Germany
Water (sewage) treatment plant	12	13
Water saving production technology	12	22
Hazardous material storage (e.g. cleaning fluids)	2	15
Other	3	6
Sample size	33	30

In Germany, as in the UK/ROI, there were twin drivers — regulation and cost. The emphasis was on using water saving technology with mixed results in terms of cost savings. The cost of installing such technology ranged from €10,000 to €90,000 with a payback of about 2-3 years. German companies also paid more attention to the methods used for storing hazardous substances than the British/Irish firms indicating the more stringent legislation in Germany.

5.5.7 Environmental Communication

In Germany environmental communication and co-operation were widely used in the textile chain (Table 5.27). The situation in the UK/ROI was rather different with few firms engaging in these initiatives. Further analysis of the German data showed that most of the compliance and compliance plus firms and all of the excellence firms were following some form of co-operative approach (Table 5.28). Co-operative approaches were made to (i) assist clients with projects (17 cases); (ii) initiate co-operative development projects with suppliers (19 cases) and (iii) to inform customers about the ecological aspects of products (17 cases). These data also showed, somewhat unexpectedly, that the Öko-Tex 100 standard was used by proportionately more compliance plus than excellence firms. There was some evidence from the interviews that excellence firms seem to be "beyond Öko-Tex 100" which was considered to be a basic requirement only.

Table 5.27. Environmental communication, UK/ROI and Germany, number of firms, FTF interviews

Initiative	UK/ROI	Germany
Environmental report/statement	3	5
Environmental product information	3	18
Öko-Tex 100 certification	2	16
Co-operational approaches	1	25
Sample size	33	30

Table 5.28. Instruments of environmental communication, Germany, % of firms, FTF

Initiative	Compliance only	Compliance plus	Excellence
Environmental report/statement	–	9.1	30.8
Environmental product information	50.0	72.7	53.8
Öko-Tex 100 certification	16.7	81.8	46.2
Co-operational approaches	66.7	72.7	100.0
Sample size	6	11	13

These initiatives were predominately market driven and although the overall economic effect was increased cost, the benefits in terms of improved market position, company image and competitiveness outweighed the additional costs.

5.5.8 Environmental Management Systems

As can be seen from the above there were certain initiatives that were more likely to be undertaken by compliance plus or excellence firms but the single initiative that differentiated the mediocre environmentally performing firms from the excellence firms was that relating to environmental management systems. Using the postal questionnaire data and allowing for the fact that firms with EMS will automatically have a higher than average number of initiatives, the difference between firms with EMS and those without is clear (Table 5.29). Despite the fact that, in all the countries, the total number of initiatives was significantly positively correlated with the number of employees there was no statistical relationship between firm size and the adoption of an environmental management system. Hence, on the basis of the postal questionnaire data, there was no evidence of an employment threshold where EMS was concerned.

Table 5.29. Average number of initiatives adopted by firms with and without an EMS, postal questionnaire

	Firms with EMS	Firms without EMS
	Average number of initiatives	Average number of initiatives
UK/ROI	6.53 (41)	3.24 (37)
Germany	9.00 (19)	5.24 (54)
Italy	3.43 (7)	2.01 (93)

Note: Figures in parenthesis denote number of firms.

Table 5.30. Average number of initiatives adopted by firms with and without an EMS, across performance group, face-to-face interviews

	Firms with EMS	Firms without EMS
	Average number of eco-points	Average number of eco-points
UK/ROI		
Compliance only	10.00 (2)	4.57 (14)
Compliance plus	15.11 (6)	9.72 (11)
Germany		
Compliance only	–	10.00 (6)
Compliance plus	16.84 (2)	15.89 (9)
Excellence	22.96 (8)	20.20 (5)
Italy		
Compliance only	–	1.50 (6)
Compliance plus	–	4.10 (20)
Excellence	9.25 (4)	6.33 (3)

Note: Figures in parenthesis denote number of firms.

Table 5.30 shows that in the UK/ROI firms that implemented certified environmental schemes such as ISO 14001, did undertake significantly more initiatives than those without certification even after allowing for the additionality effect. In Germany there was little to differentiate between the firms with certified EMS and those without in the compliance plus group. In Italy it was only the firms that were in the excellence performance category that undertook certified EMS.

However, the study was interested in not only ascertaining if a firm had an accredited EMS system but whether the firm used management techniques thereby indicating an awareness of the issues involved and the need for quality management tools such as data collection and supplier evaluation. The results for the UK/ROI and Germany are given in Tables 5.31 and 5.32 (there was no comparable data for Italy).

Table 5.31. Environmental management tools by size group, UK/ROI, FTF interviews

Size	EP	IP	ET	DC	SE	Total firms
1-10	–	–	–	–	–	1
11-25	–	1	–	2	–	3
26-50	–	–	–	3	–	5
51-100	2	2	1	3	1	6
101-250	9	3	6	8	4	11
251-500	5	3	5	6	3	7
Total	16	9	12	22	8	33

EP Environmental policy, *IP* Improvement programme, *ET* Environmental team, *DC* Data collection, *SE* Supplier evaluation.

Table 5.32. Environmental management tools by size group, Germany, FTF interviews

Size	EP	IP	ET	DC	SE	Total firms
1-10	–	–	–	–	1	3
11-25	–	–	1	2	3	4
26-50	1	1	2	4	5	6
51-100	2	1	1	4	5	5
101-250	6	6	4	7	9	9
251-500	1	2	1	3	2	3
Total	10	10	9	20	25	30

EP Environmental policy, *IP* Improvement programme, *ET* Environmental team, *DC* Data collection, *SE* Supplier evaluation.

There was a clear indication that the larger firms were more likely to make use of these environmental management tools. The results showed that although small firms may have considered the idea of adopting an environmental policy and often did collect data to use for controlling environmental improvement, they did not have a draft or an official version of the firm's environmental policy. Similarly, across the performance groups, it was the compliance plus firms in the UK/ROI and the excellence firms in Germany that had an official version of the firm environment policy or environmental improvement program. Data collection for controlling environmental improvement was undertaken by about 96% of the compliance plus firms in the UK/ROI and about 50% of the compliance only firms. In Germany none of the compliance only firms participated in data collection and the corresponding figure for the compliance plus and excellence firms that were engaged in data collection was about 85%. In the UK/ROI there was little to distinguish the compliance only and compliance plus firms in terms of environmental teams whilst in Germany it was mostly the excellence firms that had an environmental team in place. The analysis also showed that in Germany evaluation of suppliers was undertaken to some extent by 60% of compliance only firms, almost 80% of compliance plus firms and all the excellence firms. In contrast, in the UK/ROI only 15% of compliance only firms and 35% of compliance plus firms evaluated the environmental performance of suppliers.

Limited data on the initial set-up costs of an EMS were available. In the UK/ROI the median cost was £8,000 and in Germany €23,750. The motivation for firms to undertake environmental management initiatives was primarily the market in Germany and a mixture of regulation and the market in the UK/ROI. Some of the British/Irish firms acknowledged that demands from customers and venture capitalists were encouraging them to formalise their approach to environmental management systems.

5.6 Drivers for Undertaking Initiatives

Tables 5.33 (PQ) and 5.34 (FTF) give the breakdown of the dominant drivers for each initiative along with the proportion of the firms, undertaking the initiative that cited each driver. The drivers considered were regulation, cost, market (customer and supplier demands) and health and safety (H&S). In some instances not all the firms surveyed responded to this question hence the data does not sum to 100 while in other cases firms gave more than one response thus the percentages may sum to over 100.

The tables show that the predominant drivers for the initiatives undertaken corresponded reasonable well between the two surveys. Waste management and efficiency initiatives were on the whole driven by cost considerations. Initiatives related to controlling chemical treatments either through limiting their usage or replacing them with ecological alternatives reflected the importance of legislation (both environmental and statutory H&S). There was some variability between the two surveys in the importance of cost and the role of the market as drivers for change in these chemical treatment initiatives. Initiatives involving communication and co-operation with suppliers and customers as well as product certification were predominantly market driven in Germany and the UK/ROI. The Italian postal questionnaire data rather surprisingly gave H&S as the main driver for achieving certified eco-tex labelling. The drivers for the management initiatives showed some variation between countries with the UK/ROI having a regulation focus. The German results clearly demonstrated the influence of the market as a driver for organisational change and in Italy it was a combination of both market and legislation. Overall, in this industry, the role of cost and market as motivators for environmental initiatives was prominent with regulation playing a much more limiting part.

Table 5.33. Dominant drivers (cited by >20% of firms), postal questionnaire

Initiative	UK/ROI	Germany	Italy
Waste separation	Regulation 51% Cost 44%	Regulation 60% Cost 41%	Regulation 100%
Reuse of waste	Cost 80%	Cost 76% Regulation 27%	H&S 29%
Energy efficiency	Cost 80%	Cost 93%	–
Water efficiency	Cost 72% Regulation 22%	Cost 74% Regulation 20%	–
Integrated heat & water	Cost 86%	Cost 76%	–
Chemical reduction	Cost 49% Regulation 41% H&S 33%	Cost 59% Regulation 24%	–
Chemical substitution	Regulation 58% H&S 44%	Regulation 71% Market 38%	–
Process change	Cost 46% Regulation 41%	Cost 78%	–
Co-operation	Market 46% Regulation 46%	Market 58%	–
Product labelling	Market 55% Regulation 33%	Market 83%	H&S 100%
Environmental management	Regulation 66% H&S 21%	Market 74% Regulation 26% H&S 26%	H&S 100%

Table 5.34. Dominant drivers (cited by >20% of firms), face-to-face interviews

Initiative	UK/ROI	Germany	Italy
Energy efficiency	Cost 61%	Cost 92%	Cost 73%
Environmental management	Regulation 20%	Market 40%	Market 36% H&S 36%
Waste separation	Cost 75% Regulation 25%	Cost 62% Regulation 28%	Regulation 67% Cost 20%
Water use and protection	Regulation 50% Cost 32%	Regulation 52% Cost 48%	–
Chemical substitution/ reduction	Regulation 38% Cost 29% Market 24%	Market 46% H&S 23%	Market 33% H&S 33%
Packaging	Cost 40%	Market 71%	Regulation 36%
Communication	Market 20%	Market 70%	–
Air protection	Regulation 25% H&S 25%	H&S 45% Regulation 36%	–

5.7 Economic Impact of Initiatives

At interview, respondents were asked to assess the implications of undertaking each of the environmental initiatives in terms of positive, negative or neutral effects on labour, sales, costs, competitiveness, profits, price, market position and image. As the Italian dataset did not contain any detailed information on these variables for each initiative undertaken, it was only possible to calculate an aggregate figure for Italy. At the aggregate level, using all the available data, there were 190 positive or negative effects arising from 151 initiatives in the UK/ROI, 213 from 180 initiatives in Germany and 64 from 121 initiatives in Italy. These gave a ratio of effects per initiative of 1.26, 1.18 and 0.53 respectively. The ratio in the UK/ROI was slightly higher than that in Germany, which was somewhat unexpected. The very low ratio in Italy was because of the mixed nature of the sample (which included finished clothing products).

Tables 5.35 and 5.36 give a breakdown by initiative for each of the effect variables for the UK/ROI and Germany respectively. From the point of view of the firm, the overall impact of increased sales, reduced costs and associated increases in profits and competitiveness was positive. However, there was some variability in the effects between the countries for the same initiative. In the UK/ROI, cost driven initiatives such as energy efficiency, waste separation, packaging and water use reduction accounted for almost 77% of all the effects recorded. Contrary to this, in Germany the cost driven initiatives of energy efficiency, waste separation and water use contributed to about one in three of the total effects. It was the market driven initiatives of environmental management and communication along with chemical reduction and substitution that had the strongest impact in terms of effects in Germany. Environmental communications (including co-operation with suppliers and customers) made the largest contribution to the effects in Germany where in spite of the increased costs of implementing these initiatives, notable improvements in market position, image and competitiveness were recorded. Hence in Germany the market was responsible for the majority of positive impacts from the initiatives whereas in the UK/ROI cost reduction was the principal benefit. The tables show that there was little net gain in labour associated with the initiatives.

Table 5.35. Economic impact, UK/ROI, face-to-face interviews

	Labour		Cost		Price		Sales		Position		Competitive-ness		Profit		Image		Total
	↑	↓	↑	↓	↑	↓	↑	↓	↑	↓	↑	↓	↑	↓	↑	↓	
Energy efficiency				18							20		22		2		62
Environmental management	3										1	2		2	8		16
Waste separation	2			2					1		13		15		3		36
Water use & protection	1			5							7	2	8	3	4		30
Chemical reduction and/or substitution	1		1	5	1				1		6		7				22
Packaging				4					2		6		5		1	1	19
Communication									1		1				3		5
Air protection																	0
Total	7		1	34	1				5		54	4	57	5	21	1	190

Table 5.36. Economic impact, Germany, face-to-face interviews

	Labour		Cost		Price		Sales		Position		Compettive-ness		Profit		Image		Total
	↑	↓	↑	↓	↑	↓	↑	↓	↑	↓	↑	↓	↑	↓	↑	↓	
Energy efficiency	1			12	3		3		2		3		7		2		33
Environmental management	1		1	3		1	2		5		5		3		5		26
Waste separation			1	7	2		1		1		2	1	3	1	3		22
Water use and protection			4	6	1						1		3	1	1	1	18
Chemical reduction and/or substitution	1	1	6	2	3	2	2		5		5		1	2	3		33
Packaging	1		1	1									1				4
Communication	1		10	2	4	2	4		10		15	1	4	3	11	1	68
Air protection			4											2	3		9
Total	5	1	27	33	13	5	12		23		31	2	22	9	28	2	213

Table 5.37 shows the responses on the effects of the environmental initiatives obtained from the PQ. There was no comparable data for Italy hence it has been excluded from the analysis. In the UK/ROI there were 493 total effects from 388 initiatives giving a ratio of 1.27 effects per initiative and in Germany there were 719 effects from 454 initiatives, a ratio of 1.58 effects per initiative.

As before the data exhibited variability in effects for the same initiative. For example, for the product labelling initiative only 10% of the British/Irish firms that undertook this initiative indicated that this initiative had lead to an increase in costs while in Germany 50% of the firms that carried out product labelling experienced increased costs. However, this unevenness in effects was not applicable to all the initiatives undertaken. Almost equal numbers of British/Irish and German firms undertook energy efficiency and the resulting impact on reduced costs and rising profits was identical. Thus for the same driver the economic impact can differ between country.

As shown above, the experience of the firms that were interviewed was reasonably positive in terms of falling costs, increasing sales and profits and this position was mirrored by the results from the postal questionnaire. The FTF interview results illustrated the positive effect from carrying out the market driven initiative of environmental communication. A component of this group is cooperation with suppliers and customers and in the PQ this market driven initiative returned an overall slightly positive effect for Germany and a neutral effect for the British/Irish companies. This discrepancy is most likely to be due to the fact that the effects variables in the FTF dataset included non-economic measures as well as economic measures whereas in the PQ solely economic indicators were collected. The effect on the firm of environmental management initiatives was strongly negative in the UK/ROI with the majority of the firms indicating increased costs and little improvement in sales. The additional costs were almost certainly related to the extra labour required to implement the initiative.

The overall net impact on labour was substantially greater in Germany than in the UK/ROI (from Table 5.37 the total net gain in Germany was 83 and in the UK/ROI it was only 33). It was the housekeeping waste management initiatives (separation and reuse combined) as opposed to the technical or organisational initiatives that accounted for the majority of this change (67% in the UK/ROI; 60% in Germany).

One feature of this research was investigating the relationship between the environmental performance of the firm, when ranked as being compliance only, compliance plus and excellence, and the economic impact that undertaking these environmental initiatives had on the firm. Constraints within the datasets allowed this analysis to be carried out only with data from the postal survey and also excluded Italy because of the incompleteness of this country's data. When the responses from the PQ for the UK/ROI and Germany are combined, the distribution of economic effects, the total number of initiatives undertaken and the effects to initiative ratio for each environmental performance group is given in Table 5.38.

Table 5.37. Economic impact, postal questionnaire

Initiative	Sales		Labour		Costs		Profits	
	↑	↓	↑	↓	↑	↓	↑	↓
Waste separation								
UK/ROI		2	14	1	16	11	8	12
Germany	3	1	31		29	15	5	18
Reuse of waste								
UK/ROI	1	2	9		1	32	22	2
Germany	2	2	20	1	7	24	13	8
Energy efficiency								
UK/ROI	1		1		1	30	25	1
Germany	2	2	5	5	4	30	25	1
Water efficiency								
UK/ROI	2		3	1	6	16	15	3
Germany			8	3	3	29	17	2
Integrated heat & water								
UK/ROI			1		2	15	14	1
Germany		1	4	3	2	10	6	1
Chemical reduction								
UK/ROI	4	1		1	10	14	13	6
Germany	6		7	3	5	27	16	5
Chemical substitution								
UK/ROI	4		2	1	15	2	4	12
Germany	14	6	10	4	27	3	4	19
Process change								
UK/ROI	5	1	4	4	14	9	8	9
Germany	7	1	7	5	8	19	16	3
Co-operation								
UK/ROI	6		2		6	2	5	7
Germany	11		7	1	8	5	6	7
Product labelling								
UK/ROI	1				1			1
Germany	19	1	12	3	20	2	4	13
Env. management								
UK/ROI	2		7	2	20	4	2	11
Germany	7		4	4	10	2	3	2
Total Impact								
UK/ROI	26	6	43	10	92	135	116	65
Germany	71	14	115	32	123	166	115	83

Table 5.38. Economic impact, performance groups and size, UK/ROI and Germany, PQ

	Sales		Labour		Costs		Profits		TE	TI	E/I
	↑	↓	↑	↓	↑	↓	↑	↓			
C - only	–	–	3	–	5	9	7	4	28	34	0.82
C - plus	11	5	24	3	51	66	53	30	243	193	1.26
Ex.	86	15	131	39	159	226	171	114	941	615	1.53
1-25	22	12	58	12	66	71	44	52	337	245	1.38
26-50	18	2	16	9	34	41	36	27	183	127	1.44
51-100	23	2	21	8	39	62	61	24	240	151	1.59
> 100	28	4	60	12	74	122	89	43	432	301	1.44
Total	97	20	158	42	215	301	231	148	1212	842	1.44

C - only Compliance only, *C - plus* Compliance plus, *Ex.* Excellence, *TE* Total effects, *TI* Total initiatives, *E/I* Effects per initiatives.

The final column in the table shows that the ratio of effects to initiatives rises as participation in environmental initiatives increases i.e. the progression from compliance only status to compliance plus and finally excellence. This finding was in contrast to that obtained in the furniture sector where it was shown that the ratio of effects to initiatives decreased as environmental performance increased (see Table 4.26). The explanation for this discrepancy was because of the differences in the nature of the initiatives undertaken in the two sectors. In the furniture sector only 3 of the 11 initiatives were mainly cost driven (Table 4.22) whereas in the textile finishing sector cost was the predominant driver for 7 of the 11 initiatives (Table 5.33). Cost driven activities will have more obvious economic effects on firms.

From the firm's viewpoint a positive economic effect would be increased sales, reduced costs and consequently rising profits. Recalculating the ratio of effects to initiatives to include only these factors shows that the ratio for compliance only firms was 0.47; for compliance plus the ratio was 0.67 and 0.79 for excellence firms. This would indicate that compliance only firms were not capable of achieving the same economic benefits as the better environmental performing firms i.e. there was an economic incentive to adopt more initiatives.

In terms of a relationship between the size of the firm and the economic impact of environmental initiatives Table 5.38 shows that at the aggregate level there does not appear to be a clear link. However, if the positive economic effects only are considered then the very small firms did experience fewer economic benefits per initiative than the medium to large sized firms.

5.8 Obstacles to Adoption of Environmental Initiatives

At the FTF interviews the firm managers were asked to rank in order of importance the factors that hampered the introduction of environmental initiatives into their organisation. Table 5.39 shows the ranking by country for each of these factors. The ranking is based on the frequency distribution of answers with a "1" indicating the most frequently cited problem, a "2" the next most frequent answer and so on.

Table 5.39. Ranking of obstacles to adoption, face-to-face interviews

Problem	UK/ROI	Germany	Italy
Difficult to raise required capital	1	1	1
Payback period on clean technology is too long	3	2	6
Shortage of required skills and expertise	3	6	3
Clean technology is risky and unproven	–	5	8
Management priorities are focused on other issues	3	10	2
Lack of management time	2	–	–
Profit motive supersedes environmental initiatives	3	8	5
Lack of good quality advice	6	9	3
Lack of help from suppliers	6	–	4
Environmental consultancy services are expensive	–	7	7
Regulations are too uncertain to plan for new technology	4	3	–
Regulation does not encourage clean technology solutions	5	4	3

In terms of the constraints to implementing environmental initiatives there was little agreement across all the countries. The exception to this was the difficulty faced in raising capital for a new investment, which was ranked in first position by all the countries. In the UK/ROI and Germany the concerns about the payback time on clean technology were considered important but not viewed as a major constraint in Italy. Shortage of the correct skills and expertise was an issue in Italy and the UK/ROI but to a much lesser extent a concern in Germany. Italian and British/Irish firm managers also considered that from a management point of view there were too many other priorities and hence environmental issues came low down the list of concerns. Furthermore, the British/Irish firms believed that the shortage of management time to address such issues was a primary constraint. The latter was not rated as being a problem in either of the other two countries. Italian companies were concerned about the lack of good quality advice from both agencies and suppliers. The perception that regulation did not encourage firms to adopt clean technology was a concern in all the countries and the uncertainty of regulation in planning for technology changes was problematic in Germany and in the UK/ROI.

A breakdown of these obstacles by environmental performance group is given in Table 5.40. In the UK/ROI, there was little to distinguish between the two performance groups. Both groups were concerned with finding capital for investment, management issues, and skill shortages although the compliance only firms were more anxious about the risky nature of regulation. The German responses by performance group showed little difference between the groups. Capital and payback were equally important, as was the uncertainty of environmental legislation in planning for the future. Only the excellence group of firms were not concerned about introducing clean technology despite its perceived risky nature. Italian firms did show some variability between the groups especially for payback, profit and the cost of environmental consultancy services.

There was no association between the size of the firm and the ranking of the obstacles, small and large firms alike reflected the general pattern.

Table 5.40. Ranking of obstacles to adoption by environmental performance groups, face-to-face interviews

Problem	UK/ROI		Germany			Italy		
	C	C+	C	C+	E	C	C+	E
Difficult to raise required capital	1	1	1	1	1	2	1	1
Payback period on clean technology is too long	4	3	3	2	1	2	5	3
Shortage of required skills and expertise	3	4	–	3	5	2	3	1
Clean technology is risky and unproven	–	–	2	3	3	–	6	3
Management priorities are focused on other issues	4	3	–	5	–	2	1	2
Lack of management time	2	2	–	–	–	–	–	–
Profit motive supersedes environmental initiatives	2	5	–	4	6	1	5	3
Lack of good quality advice	4	–	3	–	6	–	2	1
Lack of help from suppliers	4	–	–	–	–	–	4	2
Environmental consultancy services are expensive	–	–	3	3	–	–	6	2
Regulations are too uncertain to plan for new technology	3	6	2	3	2	–	–	–
Regulation does not encourage clean technology solutions	3	–	2	3	4	–	1	3

C Compliance only, *C+* Compliance plus, *E* Excellence.

5.9 Environmental Strategy

Company strategy as it relates to environmental issues was considered to be instrumental in whether a company implemented clean technology. In all the countries the majority of firms that responded to the postal questionnaire viewed complying with regulations as the principal focus of the environmental strategy (Table 5.41, note that some firms indicated two main strategies). There was however variation between the countries in the proportion of firms adopting this approach. Italian firms, irrespective of environmental performance group, showed a widespread adoption of a regulation strategy. German firms had the lowest proportion of firms endorsing a regulation strategy and a cost strategy was considered to be almost as important. This was not unexpected considering that many of the clean technology initiatives were adopted for cost reasons. In addition, in Germany there was reasonable support for the market driven, proactive environmental strategies of promoting eco-practices and eco-products. In the UK/ROI, complying with regulations was the prime strategy but it was combined with cost savings and the promotion of eco-practices to give a mixed approach. None of the countries showed any evidence of differences in the strategy pursued by either company size or environmental performance group.

Table 5.41. Firm environmental strategy, number and percentage (in brackets) of firms, postal questionnaire

Main focus	Compliance only	Compliance plus	Excellence	All firms
UK/ROI				
Regulation	11 (73.3)	21 (72.4)	24 (70.6)	56 (71.8)
Cost savings	4 (26.7)	10 (34.5)	9 (26.5)	23 (29.5)
Promoting brand by eco-practice	–	6 (20.7)	13 (38.2)	19 (24.4)
New markets for eco-products	1 (6.7)	1 (3.4)	4 (11.8)	6 (7.7)
Promoting eco-products	–	2 (6.9)	3 (8.8)	5 (6.4)
Total firms	15	29	34	78
Germany				
Regulation	5 (55.5)	7 (38.9)	31 (67.4)	43 (58.9)
Cost savings	3 (33.3)	10 (55.5)	25 (54.3)	38 (52.1)
Promoting brand by eco-practice	1 (11.1)	4 (22.2)	10 (21.7)	15 (20.5)
New markets for eco-products	1 (11.1)	1 (5.6)	7 (15.2)	9 (12.3)
Promoting eco-products	1 (11.1)	4 (22.2)	16 (34.7)	21 (28.6)
Total firms	9	18	46	73
Italy				
Regulation	51 (76.1)	16 (50.0)	–	67 (67.0)
Cost savings	3 (4.5)	5 (15.6)	–	8 (8.0)
Promoting brand by eco-practice	–	5 (15.6)	–	5 (5.0)
New markets for eco-products	–	–	–	–
Promoting eco-products	2 (3.0)	4 (12.5)	1 (100.0)	7 (7.0)
Total firms	67	32	1	100

More than one answer was possible to this question hence the sum of the answers may exceed the total number of firms (and 100%).

Table 5.42 shows the distribution of answers to a corresponding question that asked about the factor that had the most influence on the final decision as regards environmental issues. The results for all the countries showed a fairly mixed picture. Regulation was important but to a much lesser extent than had been the case when overall company strategy was examined. Management awareness was the second most important factor in Italy and the UK/ROI, and in Germany it was as important as customer demands. In a sector where cost was the main driver for many of the initiatives it was unusual that cost as the main influential factor did not feature more strongly. When the data was tested for differences between environmental performance groups a significant difference (at $\alpha = 0.05$) was found in the German figures and in Italy between the compliance only and compliance plus there was a also a significant difference (at $\alpha = 0.01$). In both countries, the compliance only firms showed an emphasis on regulation whereas companies in the other performance groups indicated a variety of factors. There was no relationship between firm size and the factor that had the most influence in the final decision making process in any of the countries.

Table 5.42. Influences on final decision regarding environmental issues, number and percentage (in brackets) of firms, postal questionnaire

Main influence	Compliance only	Compliance plus	Excellence	All firms
UK/ROI				
Regulation	11 (73.3)	16 (55.2)	23 (67.6)	50 (64.1)
Management awareness	2 (13.3)	4 (13.8)	6 (17.6)	12 (15.4)
Cost	1 (6.7)	6 (20.7)	2 (5.9)	9 (11.5)
Customer demands	1 (6.7)	3 (10.3)	3 (8.8)	7 (9.0)
Total firms	15	29	34	78
Germany				
Regulation	5 (83.3)	4 (25.0)	20 (50.0)	29 (46.7)
Management awareness	–	6 (37.5)	6 (15.0)	12 (19.4)
Cost	–	2 (12.5)	5 (12.5)	7 (11.3)
Customer demands	1 (16.7)	4 (25.0)	9 (22.5)	14 (22.6)
Total firms	6	16	40	62
Italy				
Regulation	49 (83.0)	12 (38.7)	–	61 (67.0)
Management awareness	7 (11.9)	13 (41.9)	1 (100.0)	21 (23.1)
Cost	–	1 (3.2)	–	1 (1.1)
Customer demands	3 (5.1)	5 (16.1)	–	8 (8.8)
Total firms	59	31	1	91

Although it has been shown that there were economic benefits to be obtained from undertaking environmental initiatives, these gains did tend to be mostly associated with initiatives that were driven by cost and to a lesser extent the market. On the whole, regulation driven initiatives tended to be comparatively costly and were associated with reduced profits. Consequently, Table 5.43 shows that in both the UK/ROI and Germany few firms considered the outcomes of compliance as one whereby productivity would be improved. Almost 40% of the firms in the UK/ROI and 62% of the firms in Germany judged complying with legislation as bringing about a reduction in competitiveness. In Italy the reverse was indicated i.e. compliance led to improved productivity. There was a significant difference (at $\alpha = 0.05$) between the environmental performance groups in Germany with a high proportion of compliance only firms supporting the "good of society" view.

Table 5.43. Perception of the main impact of environmental legislation, number and percentage (in brackets) of firms, postal questionnaire

	Compliance only	Compliance plus	Excellence	All firms
UK/ROI				
Reduced competitiveness	8 (53.3)	11 (37.9)	11 (32.4)	30 (38.5)
Improvement for the good of society	5 (33.3)	11 (37.9)	15 (44.1)	31 (39.7)
Improved productivity	2 (13.3)	7 (24.1)	8 (23.5)	17 (21.8)
Total firms	15	29	34	78
Germany				
Reduced competitiveness	2 (22.2)	13 (76.5)	26 (65.0)	41 (62.1)
Improvement for the good of society	4 (44.4)	3 (17.6)	10 (25.0)	17 (25.8)
Improved productivity	3 (33.3)	1 (5.9)	4 (10.0)	8 (12.1)
Total firms	9	17	40	66
Italy				
Reduced competitiveness	5 (8.5)	5 (16.7)	–	10 (11.1)
Improvement for the good of society	36 (61.0)	13 (43.3)	1 (100.0)	50 (55.5)
Improved productivity	18 (30.5)	12 (40.0)	–	30 (33.3)
Total firms	59	30	3	90

Table 5.44 shows the responses from the FTF interviews regarding the environmental strategy adopted. In the case of Italy the conclusions were broadly the same as for the PQ results i.e. a high proportion of firms had endorsed regulation as their environmental strategy with little emphasis on environmentally proactive strategies. In the UK/ROI the company environmental strategy was almost equally divided between regulation and cost savings. Proactive strategies were virtually absent from the responses in contrast to the PQ where there was a tendency towards eco-practices particularly among the excellence firms. The German results very much confirmed the pattern from the postal questionnaire data with a mix of strategies. There was some limited evidence that when German firms move beyond compliance they adopted environmental strategies that were proactive in terms of improving the firm's market position by differentiating themselves from

their competitors through eco-practices or eco-products. However, these differences between compliance and beyond compliance firms were not statistically significant.

Table 5.44. Environmental strategy, number and percentage (in brackets) of firms, FTF

	Compliance only	Compliance plus	Excellence	All firms
UK/ROI				
Regulation	7 (46.7)	7 (41.2)		14 (43.8)
Cost saving	6 (40.0)	8 (47.1)		14 (43.8)
Eco-practices	1 (6.7)	2 (11.8)		3 (9.4)
Eco-products	1 (6.7)	–		1 (3.1)
Total firms	15	17		32
Germany				
Regulation	2 (40.0)	6 (54.5)	1 (7.7)	9 (31.0)
Cost saving	2 (40.0)	3 (27.2)	5 (38.5)	10 (34.5)
Eco-practices	–	–	6 (46.2)	6 (20.7)
Eco-products	1 (20.0)	2 (18.2)	1 (7.7)	4 (13.8)
New markets for eco-products	–	–	–	–
Total firms	5	11	13	29
Italy				
Regulation	2 (66.7)	14 (70.0)	3 (42.8)	19 (63.3)
Cost saving	1 (33.3)	4 (20.0)	–	5 (16.7)
Eco-practices	–	1 (5.0)	2 (28.6)	3 (10.0)
Eco-products	–	–	1 (14.3)	1 (3.3)
New markets for eco-products	–	1 (5.0)	1 (14.3)	2 (6.7)
Total firms	3	20	7	30

5.10 Competitive Strategy

Questions on the competitive strategy of the company were asked at the FTF interviews. Firms were categorised as belonging to one of the following groups depending on their competitive strategy: cost leadership with an emphasis on low cost, differentiation where the emphasis was on uniqueness, or niche with a focus on a particular segment of the market. Table 5.45 shows the resulting distribution. In the UK/ROI of the firms interviewed, almost half adopted a differentiation strategy with roughly equal proportions following either a cost or niche competitive strategy. In Italy the picture was fairly similar to the distribution among the British/Irish firms. In Germany the situation was somewhat different with only one firm pursuing a low cost competitive strategy and equal proportions adopting differentiation or niche strategies. *A priori*, it was anticipated that the better environmental performing firms would be clustered in either the differentiation or niche category but this was not the case for any of the countries. In fact, there was no difference between environmental performance groups and the type of strategy adopted. Also the expectation that the firms that pursued a cost leadership would undertake fewer initiatives than those with a differentiation or niche strategy was not proven (Table 5.46).

Table 5.45. Competitive strategy adopted by environmental performance group, number and percentage (in brackets) of firms, face-to-face interviews

	Cost leadership	Differentiation	Niche	All firms
UK/ROI				
Compliance only	5 (33.3)	7 (46.7)	3 (20.0)	15 (100.0)
Compliance plus	4 (23.5)	8 (47.1)	5 (29.4)	17 (100.0)
All firms	9 (28.1)	15 (46.8)	8 (25.0)	32 (100.0)
Germany				
Compliance only	–	3 (50.0)	3 (50.0)	6 (100.0)
Compliance plus	–	5 (45.5)	6 (54.5)	11 (100.0)
Excellence	1 (7.7)	6 (46.2)	6 (46.2)	13 (100.0)
All firms	1 (3.3)	14 (46.7)	15 (50.0)	30 (100.0)
Italy				
Compliance only	–	4 (80.0)	1 (20.0)	5 (100.0)
Compliance plus	5 (26.3)	10 (52.6)	4 (21.1)	19 (100.0)
Excellence	1 (14.3)	4 (57.1)	2 (28.6)	7 (100.0)
All firms	6 (19.4)	18 (58.1)	7 (22.6)	31 (100.0)

Table 5.46. Competitive strategy and average number of eco-points, face-to-face interviews

	UK/ROI	Germany	Italy
Cost leadership	8.18	20.67*	5.00
Differentiation	7.53	18.10	4.33
Niche	10.87	16.49	4.71

* One observation only.

When evaluating the relationship between the competitive strategy and the environmental strategy adopted by the firms there was a supposition that firms with a differentiation or niche competitive strategy would have had an environmental strategy that reflected this approach i.e. they would have implemented environmental strategies that were closely linked to environmentally friendly practices and products. The data in Table 5.47 shows that this was not the case in either the UK/ROI or Italy and in these countries environmental strategies driven by regulation and cost savings were the predominant approaches regardless of the company's competitive strategy. In Germany where there was only one firm with a cost leadership strategy, there was limited evidence that companies with a differentiation or niche strategy did embrace eco-strategies as well as those focused that on cost savings and regulation.

Table 5.47. Competitive and environmental strategy adopted, number of firms, FTF

	Cost leadership	Differentiation	Niche	Total
UK/ROI				
Regulation	3	8	3	14
Eco-efficiency	6	5	3	14
Eco-practices	–	2	1	3
Eco-products	–	–	1	1
Eco-markets	–	–	–	–
Total firms	9	15	8	32
Germany				
Regulation	–	4	5	9
Eco-efficiency	1	5	4	10
Eco-practices	–	3	3	6
Eco-products	–	2	2	4
Eco-markets	–	–	1	1
Total firms	1	14	15	30
Italy				
Regulation	3	11	4	18
Eco-efficiency	2	1	2	5
Eco-practices	–	3	–	3
Eco-products	–	1	–	1
Eco-markets	1	–	1	2
Total firms	6	16	7	29

5.11 Summary

German companies implemented the highest number of environmental initiatives per firm, followed by the British/Irish companies. Italian firms had the lowest average, although because of sampling difficulties this figure was not wholly comparable with that of the other countries. Compliance plus firms in Germany had better environmental performance than compliance plus firms in the other countries.

On average, large firms implemented more initiatives than small firms. In the UK/ROI the size variable influenced the relationship between competitiveness and environmental performance. In Germany, up-to-date equipment, employment in R&D and servicing an export market had a perceptible effect on environmental performance.

In general, as firms moved up the scale of environmental performance they graduated from solely "housekeeping" type initiatives, such as waste management, to activities that required technical expertise and eventually reached a level where organisational and market focused initiatives were considered to be on a par with the less complex activities. Waste management and efficiency initiatives were mostly driven by cost considerations. Initiatives related to controlling chemical treatments either through limiting their usage or replacing them with ecological alternatives reflected the importance of regulation along with cost and the market. Initiatives involving communication and co-operation with suppliers and customers as well as product certification were predominantly market driven.

In terms of the overall impact (economic and non-economic) of implementing initiatives, in Germany the market was responsible for the majority of positive impacts whereas in the UK/ROI cost reduction was the principal benefit. There were economic benefits to be obtained from undertaking environmental initiatives but these benefits did tend to be associated with initiatives that were driven by cost and to a lesser extent the market. Regulation driven initiatives tended to be comparatively costly and were associated with reduced profits.

The environmental strategy of the company varied between countries. In Italy regulation was the principal strategy, in the UK/ROI regulation and cost saving strategies dominated while in Germany a mix of strategies (regulation, cost saving and market) were followed. There was no indication that company competitive strategy and environmental strategy were interrelated.

6 Fruit & Vegetable Processing Industry

6.1 Introduction

The fruit and vegetable processing industry (NACE code 15.3) includes the preparation, preservation, canning, freezing and drying of fresh fruit and vegetables and the manufacture of fruit and vegetable juices. Wine production (NACE code 15.93) was also included; clearly this activity was only applicable in Germany and Italy. The postal questionnaire (PQ) was sent to 712 firms in Germany and 698 firms in the UK/ROI that had been identified as belonging to these sectors. Standard sources of information such as trade and telephone directories, business information publications and Internet databases were used to identify the population. The number of valid replies received was 108 for Germany and 87 for the UK/ROI. The Italian sample (100 replies) was based on a random sample of food processing firms in the Emilia-Romagna region and the short questionnaire was administered by telephone.

The choice of firms for the in-depth interviews (FTF) was made by careful consideration of the type of product manufactured, the size of the company and the extent to which, on the basis of information available, the firm could be categorised as compliance only or beyond compliance in terms of its environmental performance. In the UK/ROI sample there was some concentration on firms that processed potatoes and apples and in the Italian sample the emphasis was on tomatoes and wine. The German sample did not focus on a particular product or commodity. It included fruit juice manufacturers, wine producers and vegetable processors.

6.2 Firm Characteristics

The size distribution for the firms responding to the postal questionnaire and for firms that participated in the face-to-face interviews is shown in Tables 6.1 and 6.2. For the postal questionnaire the average employment for the UK/ROI was 41, for Germany 48 and for Italy the comparable figure was 38. These figures were consistent with the size structure of the industry as a whole, which is skewed towards the smaller firms. For the face-to-face sample, the average employment for the UK/ROI was 162, Germany 54 and Italy 79. In Germany there was no statistical difference in the average firm size between the PQ and FTF sample data. As can be seen from the size distribution of the firms that were interviewed in Italy and the UK/ROI (Table 6.2) there was a concentration of firms towards the top end of the distribution. This was because it was extremely difficult to find compliance plus firms that were willing to be interviewed in the 1-25 size category particularly in the UK/ROI. Consequently there was a significant difference in the average size of the firms sampled in the two surveys.

Table 6.1. Distribution of firms by size group, postal questionnaire

Size groups	UK/ROI		Germany		Italy	
	Number of firms	% of firms	Number of firms	% of firms	Number of firms	% of firms
1-10	34	39.5	52	48.6	46	46.0
11-25	20	23.3	20	18.7	20	20.0
26-50	12	14.0	12	11.2	14	14.0
51-100	11	12.8	10	9.3	9	9.0
101-250	6	7.0	8	7.5	10	10.0
251-500	3	3.5	5	4.7	1	1.0
Total	86	100.0	107	100.0	100	100.0

Table 6.2. Distribution of firms by size group, face-to-face interviews

Size groups	UK/ROI		Germany		Italy	
	Number of firms	% of firms	Number of firms	% of firms	Number of firms	% of firms
1-10	2	6.1	15	44.1	–	–
11-25	6	18.2	6	17.6	6	20.0
26-50	4	12.1	2	5.9	3	10.0
51-100	5	15.2	7	20.6	10	33.3
101-250	10	30.3	2	5.9	11	36.7
251-500	6	18.2	2	5.9	–	–
Total	33	100.0	34	100.0	30	100.0

For each of the sample countries, Table 6.3 shows the responses to the postal questionnaire with respect to environmental initiatives undertaken. Also shown are summary data for each of the environmental performance groups (see Sect. 2.2 for details of these initiatives and performance groups).

On the basis of the aggregate data, Table 6.3 shows that the German firms undertook substantially more initiatives than either the British/Irish firms or the Italian firms. This difference, between countries, in the average number of initiatives carried out was statistically significant (at $\alpha = 0.01$). The second part of Table 6.3 organises the data into the three environmental performance groups. In terms of the average number of initiatives per firm, the table shows a certain degree of similarity between the countries for firms in the compliance plus and excellence groups. Thus the compliance plus and excellence firms had the same level of environmental performance in all the countries. However for the compliance only firms there was sufficient evidence to reject the null hypothesis of equality (at α =

0.01) in the average number of initiatives undertaken between Italy and the UK/ROI, and also Italy and Germany indicating that the level of environmental performance for this category of firms was higher in Italy than in the other countries. This marginally higher average in Italy was most likely because all the compliance only firms carried out either one or two initiatives whereas in the UK/ROI almost 20% of firms in the sample did not carry out any initiatives; in Germany the comparable figure was 6%.

Table 6.3. Sample size, number of initiatives undertaken and environmental performance groups, postal questionnaire

	UK/ROI		Germany		Italy	
Sample size	87		108		100	
Total number of initiatives	330		566		212	
Average number of initiatives	3.79		5.24		2.12	
Environmental Performance Groups						
	No. of firms	Av. No. of initiatives	No. of firms	Av. No. of initiatives	No. of firms	Av. No. of initiatives
Compliance only	30	0.73	26	1.08	76	1.42
Compliance plus	33	4.03	27	4.15	22	4.18
Excellence	24	7.29	55	7.75	2	6.00

The face-to-face interviews with firms (FTF) enabled a more in-depth consideration of the extent and quality of environmental initiatives undertaken. Firms were asked detailed questions about eight initiatives that were particular to the fruit and vegetable processing industry and for each SME an eco-point score was calculated (see Sect. 2.2 for details).

Table 6.4 shows that Germany undertook the highest average number of initiatives, Italy had the next highest average but this was only slightly higher than that for the UK/ROI. The tests for differences in the average number of eco-points between countries were significant (at $\alpha = 0.01$) between the UK/ROI and Germany, and Italy and Germany. As the table shows the averages for the UK/ROI and Italy were very similar and there was no evidence of a statistical difference in the two means. Table 6.4 also shows the averages for each of the environmental performance groups by country. Within each performance group these were very similar and none of the tests on differences in the average between countries by environmental performance group were significant

A priori, there was the expectation that in terms of the average number of initiatives undertaken, which was a proxy for environmental performance, Germany would rank first, followed by the UK/ROI and finally Italy. The results from the postal questionnaire supported this hypothesis. The data from the face-to-face survey showed that while Germany did occupy the premier position, the UK/ROI and Italy were at the same level. There was no discernible difference in the average number of eco-points between these countries. In terms of the environmental performance groups, both surveys showed that across performance groups there was little to distinguish the countries from each other.

Table 6.4. Sample size, eco-points and environmental performance groups, FTF interviews

	UK/ROI		Germany		Italy	
Sample size	33		34		30	
Total number of eco-points	368		542		378	
Average number of eco-points	11.15		15.95		12.60	
Environmental Performance Groups						
	No. of firms	Av. No. of eco-points	No. of firms	Av. No. of eco-points	No. of firms	Av. No. of eco-points
Compliance only	16	7.73	5	8.40	11	8.51
Compliance plus	17	14.37	18	14.70	16	14.08
Excellence	–	–	11	21.42	3	19.65

Table 6.5. Economic performance of SMEs, face-to-face interviews

	UK/ROI		Germany		Italy	
	Number of firms	% of firms	Number of firms	% of firms	Number of firms	% of firms
Net profit						
Loss	3	10.3	2	6.1	1	3.4
Break-even	5	17.2	3	9.1	2	6.9
Profit 1-10%	18	62.1	17	51.5	19	65.5
Profit >10%	3	10.3	11	33.3	7	24.1
Total	29	100.0	33	100.0	29	100.0
Expected growth						
Reduced	2	6.1	2	5.9	8	28.6
Stable	–	–	10	29.4	20	71.4
Increased	31	93.9	22	64.7	–	–
Total	33	100.0	34	100.0	28	100.0

Respondents to the FTF interview, in general, characterised the firms' current profitability to be positive (Table 6.5). Only a small proportion of the firms in each country indicated that the business was making a loss. In Italy and the UK/ROI these loss-making firms were mostly compliance only firms whereas in Germany they were firms that had moved beyond compliance. Expectations of growth, in terms of sales, during the next three years are shown in the second part of the table. In Germany and the UK/ROI the majority of firms were predicting an expansion in sales whereas in Italy none of the firms interviewed reported a projected growth in sales. This predicted expansion in Germany was almost exclusively in the compliance plus and excellence firms.

In all the countries, the two factors that managers considered to be vital for competitiveness in the business were price and quality with service factors having secondary importance. Factors such as capital availability, marketing, the environment and consultancy services were not thought to have an influence on firm competitiveness.

6.3 Size Effect

Table 6.6 shows the relationship between firm size and environmental performance, measured as the average number of initiatives per firm for the postal questionnaire data and as the average number of eco-points per firm for the data collected at interview. In terms of the PQ data, there was a statistically significant positive linear relationship between firm size and the number of initiatives undertaken in the UK/ROI (where the correlation coefficient was 0.367 and significant at $\alpha = 0.01$) and in Germany ($r = 0.284$ at $\alpha = 0.01$). Tests for differences in the employment means between the environmental performance groups in the UK/ROI and in Germany were statistically significant hence supporting the connection between the size of the firm and environmental performance. As can be seen from Table 6.6 there was little variability in the average number of initiatives between the size groups in Italy. There was no significant relationship between firm size and environmental performance in Italy; all firms were equally likely to undertake the same number of initiatives.

Table 6.6. Average number of initiatives (PQ) and eco-points (FTF) by size group

Size groups	UK/ROI		Germany		Italy	
	PQ	FTF	PQ	FTF	PQ	FTF
1-10	2.59	12.50	3.69	14.71	1.98	–
11-25	3.80	8.72	6.10	15.39	2.25	14.26
26-50	4.50	11.00	7.08	16.00	2.21	11.66
51-100	5.18	11.53	7.00	19.05	2.56	12.57
101-250	5.50	11.00	6.63	15.00	2.00	11.96
251-500	6.67	13.16	7.80	17.00	2.00	–
Total	3.79	11.15	5.24	15.95	2.12	12.60

For the sample of firms that participated in the FTF interviews, Table 6.6 also shows the distribution of eco-points by firm size. In contrast to the postal questionnaire data, the distribution clearly shows the lack of a linear relationship between the size of the firm and environmental performance, which was confirmed by the non-significant correlation coefficients. Consequently the characteristics of the environmental performance groups can be compared without the effect of size distorting the data.

6.4 Competitive Status

As with textiles, the indicators of competitiveness collected in the postal questionnaire were firm profitability, value added per employee, employment in R&D, and employment growth in the previous five-year period and destination of sales.

In all the countries, firm profitability and value-added per employee, relative to the industry average, were highly correlated with each other (in the UK/ROI $r = 0.691$, in Germany $r = 0.598$ and in Italy $r = 0.566$; significant at $\alpha = 0.01$ in all cases). Environmental performance, as measured by the number of initiatives undertaken, was not correlated with either of these variables in any of the countries (Table 6.7). Tables 6.8 and 6.9 show that there was no difference in profitability or value added between the environmental performance groups. Thus, compliance only firms were as likely to have above average profitability and value added as the better performing firms in relative terms. Earlier it was shown that in absolute terms compliance plus firms usually performed better.

Table 6.7. Firm profitability and value added per employee, average number of initiatives, postal questionnaire

Compared to industry average	Firm Profitability	Value added per employee
UK/ROI		
Below average	3.19	3.05
Average	3.95	3.88
Above average	4.29	4.35
Germany		
Below average	4.71	5.09
Average	5.33	5.75
Above average	6.00	5.67
Italy		
Below average	1.55	1.46
Average	2.29	2.19
Above average	2.36	2.42

Table 6.8. Firm profitability number and percentage (in brackets) of firms, PQ

Compared to industry average	Compliance (0-2 inits)	Compliance + (3-5 inits)	Excellence (6-11 inits)	All firms
UK/ROI				
Below average	10 (34.5)	10 (30.0)	6 (26.1)	26 (30.6)
Average	13 (44.8)	17 (51.5)	12 (52.2)	42 (49.4)
Above average	6 (20.7)	6 (18.2)	5 (21.7)	17 (20.0)
Total firms	29	33	23	85
Germany				
Below average	6 (25.0)	6 (24.0)	9 (17.3)	21 (20.8)
Average	15 (62.5)	16 (64.0)	32 (61.5)	63 (62.4)
Above average	3 (12.5)	3 (12.0)	11 (21.2)	17 (16.8)
Total firms	24	25	52	101
Italy				
Below average	21 (28.0)	1 (4.5)	–	22 (22.2)
Average	43 (57.3)	19 (86.4)	1 (50.0)	63 (63.6)
Above average	11 (14.7)	2 (9.1)	1 (50.0)	14 (14.4)
Total firms	75	22	2	99

Table 6.9. Value added per employee, number and percentage (in brackets) of firms, PQ

Compared to industry average	Compliance (0-2 inits)	Compliance + (3-5 inits)	Excellence (6-11 inits)	All firms
UK/ROI				
Below average	8 (27.6)	7 (21.2)	4 (17.4)	19 (22.4)
Average	16 (55.2)	19 (57.6)	14 (60.9)	49 (57.6)
Above average	5 (17.2)	7 (21.2)	5 (21.7)	17 (20.0)
Total firms	29	33	23	85
Germany				
Below average	4 (22.2)	6 (24.0)	12 (22.6)	22 (22.9)
Average	10 (55.6)	17 (68.0)	29 (54.7)	56 (58.3)
Above average	4 (22.2)	2 (8.0)	12 (22.6)	18 (18.8)
Total firms	18	25	53	96
Italy				
Below average	12 (15.8)	1 (4.5)	–	13 (13.0)
Average	55 (72.4)	19 (86.4)	1 (50.0)	75 (75.0)
Above average	9 (11.8)	2 (9.1)	1 (50.0)	12 (12.0)
Total firms	76	22	2	100

The extent of R&D activity, measured as the number of full-time equivalent employees working in R&D[20], is shown in Table 6.10.

Table 6.10. R&D staff (full-time equivalents), postal questionnaire

Number of R&D staff (FTEs)	Average number of initiatives		
	UK/ROI	Germany	Italy
0	2.81	3.54	1.56
0.5-1.5	3.50	6.77	2.25
2.0-3.5	5.36	7.38	2.14
4+	7.89	8.00	2.77
Correlation coefficient (R&D FTEs and number of initiatives)	n.s.	0.347***	0.251**

*** significant at 0.01; ** significant at 0.05.

In Germany and Italy there was a significant relationship between the number of initiatives and the degree of R&D activity in the firm. Earlier in the chapter it was shown that firm size was positively related to the number of initiatives. When the measure of R&D was tested against size there was positive correlation indicating that larger firms have more R&D capacity than smaller firms (Table 6.11). Returning to the relationship between R&D and environmental performance and at the same time taking into account the effect of size, both Germany and Italy retained this relationship (respective partial correlation coefficients were $r = 0.245$ at $\alpha = 0.05$ and $r = 0.306$ at $\alpha = 0.01$). Although the British/Irish firms, on average, did undertake more R&D than the other countries there was no evidence to suggest that this lead to improved environmental performance in the firms.

Table 6.11. Employment in R&D by firm size, postal questionnaire

Size groups	Average number of FTEs in R&D		
	UK/ROI	Germany	Italy
1-10	0.46	0.15	0.87
11-25	1.25	0.71	1.68
26-50	2.17	1.04	2.00
51-100	3.32	1.63	4.33
101-250	7.13	2.00	3.45
251-500	2.17	9.00	6.00
All firms	1.64	0.95	1.08
Correlation coefficient (employment and R&D FTEs)	0.377***	0.814***	0.411***

*** significant at 0.01.

[20] A full-time equivalent means that part-time employees were counted as 0.5 while their full-time counterparts were counted as one.

Labour growth, as measured by the change in employment in the firm over the previous five-year period, varied between the countries with Germany having, on average, 17% growth; Italy 30% and the UK/ROI 25%. No statistically significant relationship with environmental performance was evident in any of the countries.

The final measure of competitiveness, from the postal questionnaire, was constructed using information based on the destination of sales. For each country, Table 6.12 gives details of the criteria, the firm distribution and the average number of initiatives undertaken. The average level of exports in both the UK/ROI and Germany was around 7%. However, in Italy the average was 28% indicating a more export-orientated approach to sales. The inclusion of wine producing firms in the Italian sample clearly contributed to this high figure. There was some experimentation using a different ranking system for Italy. The results showed that there was no significant difference in the conclusions therefore the same criteria were used in all the countries.

Table 6.12 shows that in Germany and Italy firms that exported did in fact undertake more environmental initiatives. There was also evidence that smaller firms tended to predominately serve local and national markets. When the size factor was taken into account the relationship between destination of sales and environmental performance remained significant. In Germany there was a distinct relationship between the destination of sales and environmental performance (Table 6.13). The compliance only firms mostly catered for the local and national markets whereas the excellence firms were well represented in the international market. The Italian figures in this table give a rather mixed picture and this was primarily because the vast majority (76%) of Italian firms were categorised as compliance only i.e. they carried out only one or two initiatives. The British/Irish data did not exhibit any relationship between sales destination and environmental performance.

Table 6.12. Sales destination ranking, postal questionnaire

Rank	Rule	UK/ROI		Germany		Italy	
		No. of firms	No. of inits.	No. of firms	No. of inits.	No. of firms	No. of inits.
1	Local sales only	9	2.89	34	4.79	5	2.00
2	Local and national sales only	39	3.62	33	4.39	13	1.46
3	Exports < 20%	28	3.54	21	5.90	32	1.88
4	Exports >= 20%	9	5.67	14	7.14	50	2.46
Correlation coefficient (destination and initiatives)		n.s		0.238**		0.266***	

*** significant at 0.01; ** significant at 0.05.

Table 6.13. Competitiveness ranking and environmental performance, postal questionnaire

	All firms	Compliance only	Compliance plus	Excellence
	No. (% of firms)	% of firms	% of firms	% of firms
UK/ROI				
Local sales only	9 (10.6)	55.6	22.2	22.2
Local and national sales only	39 (45.9)	41.0	28.2	30.8
Exports < 20%	28 (32.9)	28.6	57.1	14.3
Exports >= 20%	9 (10.6)	11.1	44.4	44.4
Germany				
Local sales only	34 (33.3)	29.4	29.4	41.2
Local and national sales only	33 (32.4)	30.3	30.3	39.4
Exports < 20%	21 (20.6)	19.0	19.0	61.9
Exports >= 20%	14 (13.7)	7.1	14.3	78.6
Italy				
Local sales only	5 (5.0)	80.0	–	20.0
Local and national sales only	13 (13.0)	100.0	–	–
Exports < 20%	32 (32.0)	84.4	15.6	–
Exports >= 20%	50 (50.0)	64.0	34.0	2.0

In summary, on the basis of the postal questionnaire data alone, there was some evidence in Germany and Italy to support the hypothesis that the more competitive firms undertake more environmental initiatives. Firms that participated in R&D and also served an export market were more likely to have better environmental performance irrespective of size.

In addition to data on the destinations of sales, employment growth and participation in R&D data was collected at the face-to-face interviews on other explanatory variables such as productivity per head, age of machinery, and employee qualifications. The correlation analysis of the relationship between these measures of competitiveness and environmental performance for the FTF interviews showed that there were no significant relationships in any of the countries (Table 6.14). For each country, the average value of these competitiveness measures is shown in Tables 6.15 to 6.17.

Table 6.14. Correlation between measures of competitiveness and environmental performance, face-to-face interviews

	Expected sign	Correlation coefficient	Level of significance
UK/ROI			
Exports (%)	+	-0.037	ns
Age of Machinery	-	0.024	ns
Productivity (turnover/capita)	+	0.091	ns
R&D (% of total employment)	+	0.201	ns
Growth (employment, 1993-1998)	+	0.351	ns
Qualifications (% of total employment)	+	0.028	ns
Germany			
Exports (%)	+	-0.124	ns
Age of Machinery	-	-0.120	ns
Productivity (turnover/capita)	+	0.048	ns
R&D (% of total employment)	+	-0.291	ns
Growth (employment, 1993-1998)	+	0.202	ns
Qualifications (% of total employment)	+	-0.013	ns
Italy			
Exports (%)	+	-0.330	ns
Age of Machinery	-	-0.131	ns
Productivity (turnover/capita)*	+	n.a.	
R&D (% of total employment)	+	-0.149	ns
Growth (employment, 1993-1998)	+	0.082	ns
Qualifications (% of total employment)	+	-0.007	ns

*Not available for Italy.

Table 6.15. UK/ROI: Average competitiveness measures by environmental performance groups, face-to-face interviews

	Compliance only	Compliance plus	All Firms
Av. Turnover (£ m)	12.510	16.450	14.440
Av. Productivity per head (£)	82,760	134,370	107,530
Av. Level of exports (%)	8.1	14.4	11.3
Av. Employment per firm	140	182	162
Av. Employment growth (%)	61.2	109.0	88.9
Av. Development staff per firm	1.75	2.73	2.26
Av. Qualified staff (% of all staff)	14.76	14.91	14.85
Av. Wages (£ per hour)	5.29	4.89	5.05
Av. Age of machinery (years)	5.66	6.49	6.05
Sample size	16	17	33
Average number of eco-points	7.73	14.37	11.15

Table 6.16. Germany: Average competitiveness measures by environmental performance groups, face-to-face interviews

	Compliance only	Compliance plus	Excellence	All Firms
Av. Turnover (€, million)	1.200	12.440	7.385	9.050
Av. Productivity per capita (€)	11,350	16,010	10,030	13,310
Av. Level of exports (%)	13.2	6.7	4.9	7.1
Av. Employment per firm	9	54	73	54
Av. Employment growth (%)	11.7	23.4	108.4	47.0
Av. Development staff per firm	1.00	5.05	10.9	7.35
Av. Qualified staff (% of all staff)	–	49.51	47.59	48.71
Av. Wage (€ per hour)	7.37	10.91	9.37	9.33
Av. Age of machinery (years)	10.83	5.95	7.34	7.11
Sample size	5	18	11	34
Average number of eco-points	8.40	14.70	21.42	15.95

Table 6.17. Italy: Average competitiveness measures by environmental performance groups, face-to-face interviews

	Compliance only	Compliance plus	Excellence	All Firms
Av. Turnover (€, million)	–	–	–	–
Av. Productivity per capita (€)*	–	–	–	–
Av. Level of exports (%)	52.3	32.5	53.3	41.8
Av. Employment per firm	86	67	117	79
Av. Employment growth (%)	34.5	51.6	18.7	42.0
Av. Development staff per firm	4.33	4.82	2.58	4.42
Av. Qualified staff (% of all staff)	29.3	25.1	30.4	27.2
Av. Worker (€ per hour)*	12	12	12	12
Av. Age of machinery (years)	5.94	4.66	3.98	5.06
Sample size	11	16	3	30
Average number of eco-points	8.51	14.08	19.65	12.60

* Estimates based on interviews with sector experts as opposed to firm managers.

There was no difference in the means between the environmental performance groups for any of these variables in either the British/Irish or the Italian sample data. In the German case there was a significant difference between performance groups for the age of machinery with the better environmental performing firms availing of more up-to-date equipment.

6.5 Environmental Initiatives

Figure 6.1 shows the variability between the countries in the adoption of the individual environmental initiatives by SMEs that responded to the postal questionnaire.

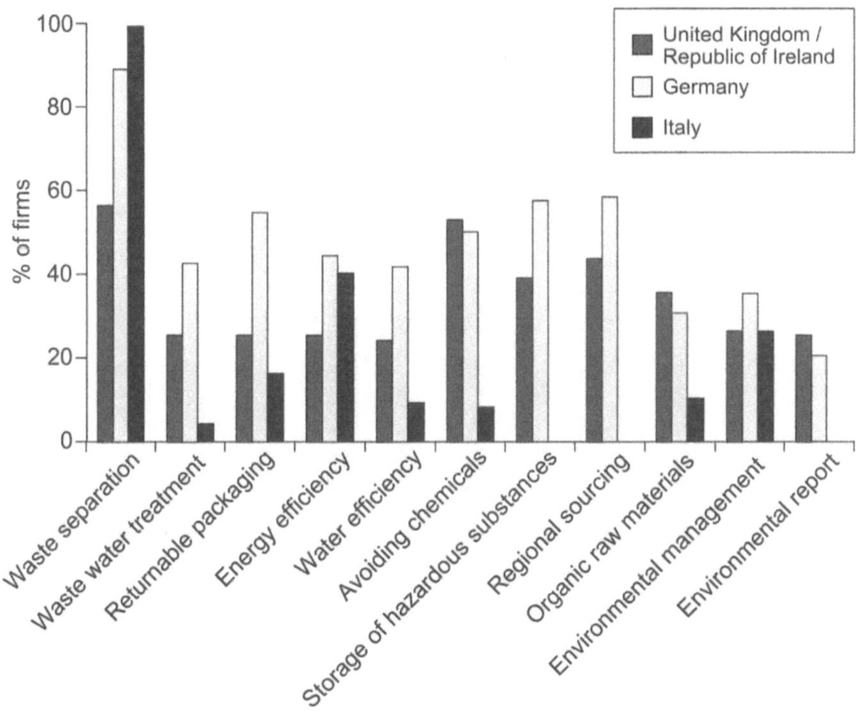

Fig. 6.1. Proportion of firms undertaking each initiative, all countries, postal questionnaire

Almost all the German and Italian firms undertook waste separation but in the UK/ROI this activity was undertaken by just over half of the companies. This was undoubtedly due to the more stringent legislation regarding waste management in both Germany and Italy. Energy and water efficiency initiatives, which were aimed at reducing firm costs, showed variation in the uptake between countries. Energy efficiency initiatives were undertaken by a significantly smaller proportion of firms in the UK/ROI than in either Germany or Italy. However, in Italy comparatively few firms undertook water efficiency initiatives. Market focused initiatives, such as regional sourcing, the use of organic raw materials and the avoidance of chemical ingredients showed no significant variation between the UK/ROI and Germany in terms of the proportion of firms that undertook these initiatives. The Italian sample showed a very small percentage of firms engaging in these customer orientated initiatives, which was unexpected given the importance of good quality food to the Italian consumer. Management initiatives that would lead to improved environmental procedures were adopted in roughly equal proportions in all the countries. On the whole, the level of participation in the various initiatives was almost always higher in the German companies. The British/Irish firms participated, to some degree, in all the initiatives whereas the Italian companies concentrated mostly on waste separation and energy efficiency.

The distribution, for each country, of the firms undertaking each initiative in the three environmental performance groups (i.e. compliance only, compliance plus and excellence) is shown Figures 6.2 to 6.4.

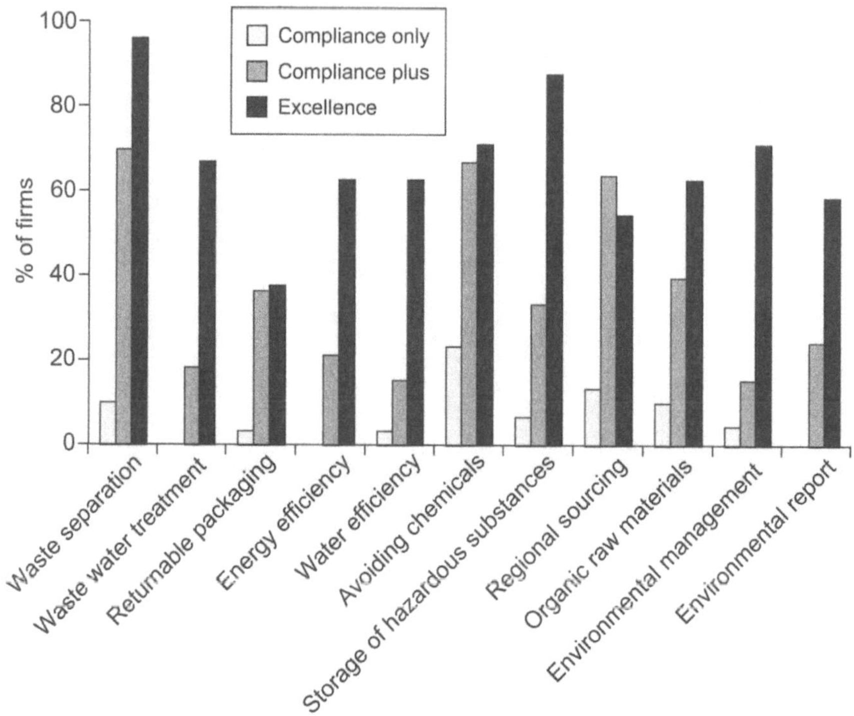

Fig. 6.2. Proportion of firms undertaking each initiative, UK/ROI, postal questionnaire

In the furniture and textile finishing industries, compliance only firms in all the countries were shown to concentrate on waste management initiatives with little participation in the market driven initiatives. In the fruit and vegetable processing industry there was a distinct difference between countries in the type of initiative that compliance only firms undertook. In the British/Irish sample this category of firms focused on market driven initiatives such as avoiding chemical ingredients, regional sourcing and organic raw materials. In contrast in the German sample, compliance only firms were engaged in mostly waste separation and packaging with limited participation in market initiatives. Italian compliance only firms chiefly undertook waste separation and energy efficiency initiatives. A common feature of the compliance only firms in all the countries was the lack of participation in environmental management initiatives. This was also a facet of both the furniture and textile finishing sectors

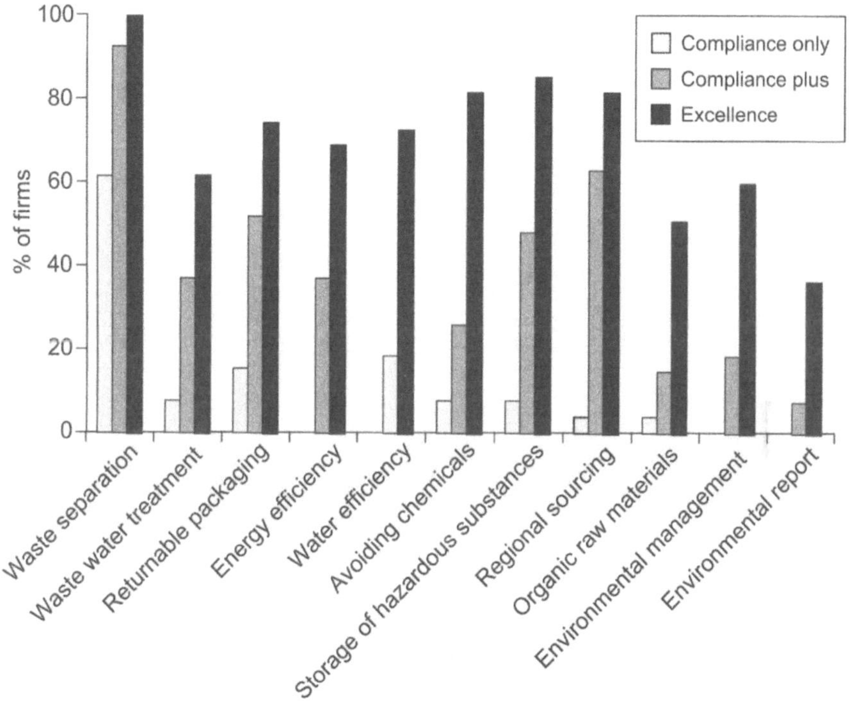

Fig. 6.3. Proportion of firms undertaking each initiative, Germany, postal questionnaire

Compliance plus firms in all the countries were engaged in a more extensive range of initiatives but to a much lesser extent than the excellence performing firms. In the UK/ROI the features that distinguished the compliance plus firms from the excellence firms were that compliance plus firms carried out fewer efficiency initiatives, had minimal participation in waste water treatment and limited take-up of management initiatives. In Germany compliance plus firms lacked involvement in the market driven and management initiatives and to some extent initiatives associated with water (both treatment and efficiency). Italian compliance plus differed from the excellence firms in that proportionately fewer compliance plus firms undertook water efficiency initiatives and the market driven initiatives associated with artificial and organic materials.

As was shown in Section 6.3, there was a relationship between the size of the firm and the number of initiatives undertaken in the UK/ROI and Germany. In the UK/ROI there was no statistical difference in the size of the firm and the ability to undertake the market driven initiatives. Small firms were as likely as large firms to introduce initiatives to limit the use of artificial ingredients, and to increase the usage of organic raw materials and regional sourcing. As size increased proportionately more firms undertook efficiency and management initiatives. In Germany, as firm size increased proportionately more firms were involved in management and market initiatives. Small firms in Germany concentrated on waste and packaging initiatives. In Italy there was no size effect associated with the type of initiatives undertaken.

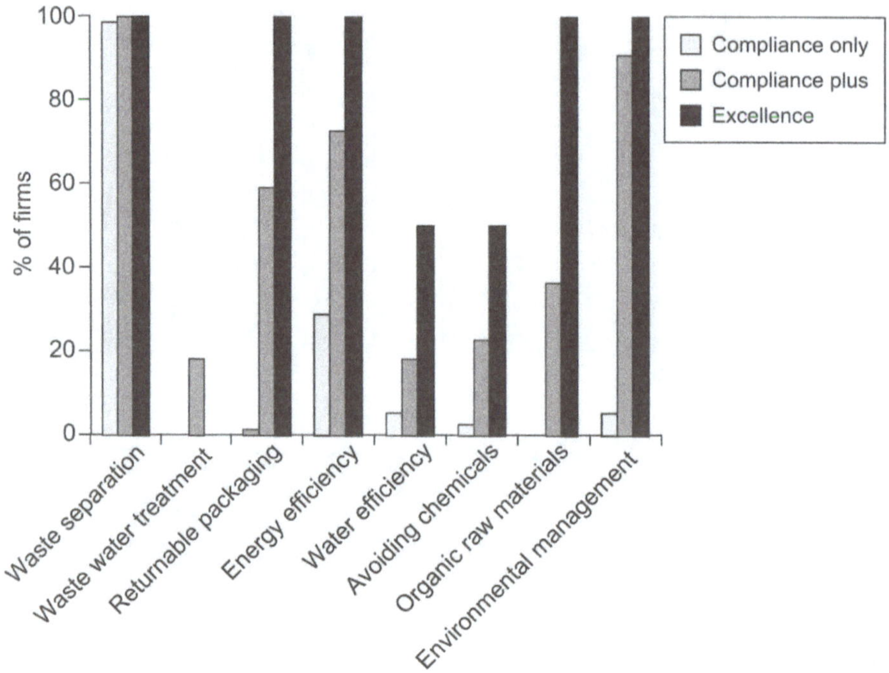

Fig. 6.4. Proportion of firms undertaking each initiative, Italy, postal questionnaire

In the UK/ROI, the data from the face-to-face interviews (Table 6.18) did not follow the pattern of SMEs sampled in the postal survey. Compliance only firms did not show the level of involvement in the market initiatives (particularly the usage of organic raw materials and regional sourcing) relative to the other initiatives. There was a high degree of adoption of most initiatives irrespective of performance group and there was no significant difference in the take up of waste separation, packaging, energy efficiency, or environmental management initiatives between the performance groups. The initiatives that most clearly differentiated the compliance plus firms from the compliance only firms were environmental communication, water use and protection, and as referred to above, the market initiatives.

Table 6.18. UK/ROI: Environmental initiatives by performance groups, FTF interviews

Initiative	Compliance only		Compliance plus		All firms	
	No. of firms	%	No. of firms	%	No. of firms	%
Waste reduction	14	87.5	17	100.0	31	93.9
Packaging	14	87.5	16	94.1	30	90.9
Water use and protection	12	75.0	16	94.1	28	84.8
Energy	12	75.0	15	88.2	27	81.8
Avoiding artificial ingredients	11	68.8	15	88.2	26	78.8
Organic raw materials and regional sourcing	6	37.5	15	88.2	21	63.6
Environmental management	13	81.3	17	100.0	30	90.9
Environmental communication	2	12.5	11	64.7	13	39.4
Sample size	16		17		33	

Table 6.19. Germany: Environmental initiatives by performance groups, face-to-face interviews (figures in brackets are % of sample)

Initiative	Compliance only	Compliance plus	Excellence	All firms
Waste reduction[a]	2 (40.0)	8 (44.4)	7 (63.6)	17 (50.0)
Packaging	4 (80.0)	16 (88.9)	10 (90.9)	30 (88.2)
Water use and protection	3 (60.0)	17 (94.4)	11 (100.0)	31 (91.2)
Energy	2 (40.0)	13 (72.2)	8 (72.7)	23 (67.6)
Avoiding artificial ingredients	5 (100.0)	17 (94.4)	10 (90.0)	32 (94.1)
Organic raw materials and regional sourcing	–	12 (66.7)	11 (100.0)	23 (67.6)
Environmental management[b]	–	1 (5.6)	5 (45.5)	6 (17.7)
Environmental communication	3 (60.0)	12 (66.7)	11 (100.0)	26 (76.5)
Sample size	5	18	11	34

[a] Waste separation as an initiative if five or more waste streams were separated. These were household type waste, organic waste, paper and cardboard, PE-film and glass. These types of waste are usually separated in German households.
[b] Only companies that had implemented a management system such as ISO14001 were included.

Table 6.19 shows the distribution of the initiatives undertaken by the firms in Germany that took part in the interviews separated into the three performance groups. As with the PQ results the excellence performance firms were distinguishable from the other performance groups by the fact that they had a high degree of participation in the management initiatives, EMS and communications, and the market driven initiatives especially organic raw materials and regional sourcing. As regards the other initiatives, there was no significant difference in the take up by performance group. Waste, packaging, energy and water initiatives were as likely to be undertaken by the compliance only as by the compliance plus firms.

The results from the Italian firms interviewed (Table 6.20) showed that there was a significant difference between compliance only and compliance plus firms in terms of energy efficiency, environmental communications and usage of organic materials. These differences were also observed in the postal questionnaire data.

Table 6.20. Italy: Environmental initiatives by performance groups, face-to-face interviews (figures in brackets are % of sample)

Initiative	Compliance only	Compliance plus	Excellence	All firms
Waste reduction	10 (90.9)	16 (100.0)	3 (100.0)	29 (96.7)
Packaging	11 (100.0)	16 (100.0)	3 (100.0)	30 (100.0)
Water use and protection	10 (90.9)	15 (93.8)	3 (100.0)	28 (93.3)
Energy	5 (45.5)	14 (87.5)	3 (100.0)	22 (73.3)
Avoiding artificial ingredients	11 (100.0)	16 (100.0)	3 (100.0)	30 (100.0)
Organic raw materials and regional sourcing	–	7 (43.8)	3 (100.0)	10 (33.3)
Environmental management	8 (72.7)	16 (100.0)	3 (100.0)	27 (90.0)
Environmental communication	2 (18.2)	7 (43.8)	3 (100.0)	12 (40.0)
Sample size	11	16	3	30

Detailed information on the individual environmental initiatives was collected during the face-to-face interviews. The remainder of this section presents the findings.

6.5.1 Waste Reduction

Most of the companies interviewed undertook some form of waste separation although the coverage varied between countries. Table 6.21 shows, for each country, the various waste streams that were separated and the distribution of firms participating. All of the German firms interviewed separated the waste produced and a high proportion of waste went for recycling rather than disposal. In all the countries, organic waste was commonly disposed of to local farmers for use as fertlizer or animal feed. Although in Germany one innovative firm had linked up with a local gas company and supplied organic waste for the production of gas while another company was investigating the possibility of installing a Biogas production unit on site which would lead to savings on waste, water and heating. The latter would require an investment of around €1 million with an extensive payback period. Paper and cardboard were often incinerated or recycled along with PE-film. Recycling has frequently involved the company making an investment in a compactor but most firms viewed this as a sound investment that will produce cost savings in the long term. In contrast to the situation in Germany and Italy, few of the British/Irish firms recycled glass waste. This was an indication of the lack of facilities for glass recycling in the UK/ROI.

Table 6.21. Waste management, number of firms, face-to-face interviews

Waste stream	UK/ROI	Germany	Italy
Organic waste	23	33	30
Paper and cardboard	29	29	22
PE-film	19	29	22
Glass	5	22	24
Steel or tins	13	18	21
Waste chemicals (special waste)	10	4	16
Filter cake or sludge	16	13	9
Household type waste	25	22	27
Other	-	3	2
Sample size	33	34	30

The impetus to separate waste was a mixture of cost and regulation. Firms recognised a need to reduce the amount of waste that was sent to landfill particularly in the UK/ROI where such charges have been progressively rising. The British/Irish and German data showed that for each of these waste streams there was a net cost effect particularly regarding the separation and disposal of PE film. However, these were offset by the need to comply with legislation and reduce the landfill costs.

Limited data relating to the level of financial investment involved to facilitate waste separation was available only for some of the German sample firms. Amounts invested averaged around €20,000 for the purchase of automatic separators and compactors.

6.5.2 Energy Initiatives

Table 6.22 shows the implementation of the different energy efficient initiatives available. There was some variation across country with the British/Irish and Italian firms concentrating on the installation of the relatively inexpensive energy efficient lighting, which has led to low savings on the overall energy bill. More sophisticated systems of energy reduction did require substantial capital investments. For example in Germany, the cost of installing an integrated heat and water system ranged from €50,000 to €2.5 million and in the UK/ROI such systems have required investments of up to £400,000. The payback period for such investments was, on average, less than 5 years. Such technology was installed in the mid to late 1990s and in many cases firms were already experiencing benefits in terms of cost savings, which were, on the whole, rated as high. These more complex systems were usually operational in firms that were considered as beyond compliance and in the UK/ROI they were generally found in firms that employed more than 100 workers.

Table 6.22. Energy initiatives, number of firms, face-to-face interviews

Initiative	UK/ROI	Germany	Italy
Efficient heat & power production e.g. combined heat and power plant	3	8	20
Efficient energy distribution	6	2	1
Efficient energy use or re-use e.g. integrated heat and water management	8	12	1
Energy efficient lighting	26	4	22
Other	4	11	–
Sample size	33	34	30

One interesting development in Germany has been the introduction of solar collectors in a few companies. The capital investment required was usually under €9,000 but the outcomes, in terms of savings on energy costs, have not yet been quantified as most of these are fairly recent installations.

The main motivation for introducing energy efficient initiatives was to achieve cost savings and firms that had invested in such initiatives did acknowledge a reduction in costs and hence increased profitability.

6.5.3 Packaging

Packaging was regarded as an issue that was fundamental to SMEs in the food processing industry. In the UK/ROI and Germany the main reason for using environmentally friendly packaging was the demands of customers whilst in Italy this initiative was very much regulation driven. In Germany and Italy the main type of packaging employed was returnable packaging, one-way glass packaging and polyethylene film. In the UK/ROI there was little demand for one-way glass packaging and the emphasis was on PE-film and to a much lesser extent returnable packaging (Table 6.23).

In Germany there was wide variation in the capital investments made in packaging initiatives during the late 1990s. These ranged from €122,000 to €3.25 million while in the UK/ROI the average investment on packaging equipment was about £100,000. There was no data on the economic impact of these specific packaging initiatives but in the interviews firms did acknowledge that they had contributed to an increase in competitiveness.

Table 6.23. Packaging initiatives, number of firms, face-to-face interviews

Initiative	UK/ROI	Germany	Italy
Returnable packaging e.g. glass bottles	12	24	17
One way glass packaging	6	15	15
Polyethylene film	21	15	12
Tetra pack	–	2	4
Tins	3	1	12
Other	13	7	7
Sample size	33	34	30

6.5.4 Avoiding Artificial Ingredients

In all the countries, customer demands was the main motivator for avoiding artificial ingredients in the finished product. In addition, in Germany there was legislation prohibiting the use of colourings, flavourings and preservatives. Table 6.24 shows that in Germany there was virtually universal avoidance of artificial ingredients. In Italy the situation was similar but in the UK/ROI, with the exception of genetically modified ingredients, there was more limited take up of this initiative.

Table 6.24. Avoidance of artificial ingredients, number of firms, face-to-face interviews

Initiative	UK/ROI	Germany	Italy
Artificial colours	17	31	26
Preservatives	10	24	20
Genetically modified organisms or substances which are produced by genetic technology	25	32	28
Artificial aromas or flavour	16	29	20
Taste intensifying substances	14	29	17
Other	–	–	3
Sample size	33	34	30

There was a distinct difference in the uptake of these initiatives by environmental performance group with the compliance only firms undertaking substantially fewer. There was no indication of a size effect in either the UK/ROI or Germany. The dominant driver for undertaking these initiatives was regulation in the UK/ROI and market pressures in Germany. In general, controlling chemical inputs was costly and the firms did not on the whole achieve any direct cost benefit but were able to record an increase in market position and image.

6.5.5 Organic Raw Materials and Regional Sourcing

Organic raw materials were used most extensively in the German firms (Table 6.25). The proportion of organic material inputs to total inputs varied between firms within and across countries. In the UK/ROI, among the firms involved in using organic raw materials the highest proportion cited was 10% of total raw inputs. In contrast, in Germany some of the firms were using 100% organic inputs. At the interviews in the UK/ROI and Germany, managers drew attention to the cost of organic inputs, often three times the price of non-organic raw materials, and that although there was a demand from the market for organic inputs the customer was not always willing to pay the increased price of the organic alternative. Of the 19 companies in Germany that used organic products, 11 indicated that there had been an associated cost increase. There was some evidence of regional sourcing of inputs and in many cases the source of the raw materials were local farmers. The general view among managers in the UK/ROI and Germany was that these initiatives led to increased sales and improved company image.

Table 6.25. Organic raw materials, number of firms, face-to-face interviews

Initiative	UK/ROI	Germany	Italy
Products of organic agriculture	14	19	10
Products produced under contracts containing elements of organic agriculture (e.g. integrated farming in Germany)	1	6	–
Others	1	–	–
Sample size	33	34	30

6.5.6 Water Use and Protection

The fruit and vegetable processing industry uses water extensively in the cleaning of the raw materials and where applicable the washing of returnable glassware. As such this was an area where firms were trying to minimise costs. In addition, industrial water use was highly regulated thereby requiring the treatment of waste water effluent. In all countries, initiatives to reduce water consumption and treat waste were fairly common (Table 6.26).

Table 6.26. Water use and protection, number of firms, face-to-face interviews

Initiative	UK/ROI	Germany	Italy
Water (sewage) treatment plant	12	11	19
Water saving production technology	15	19	13
Hazardous material storage (e.g. cleaning fluids)	18	13	4
Other	1	-	-
Sample size	33	34	30

Investment in water treatment plants averaged between €0.5 and €0.75 million. These plants were in large (over 100 employees) compliance plus and excellence firms. Comparable data from the British/Irish sample was £1.5 million and these water treatment plants were almost exclusively confined to large, compliance plus firms. Water treatment plants not only had substantial capital costs but also, in some firms, had high running costs and required additional technical staff. Water saving technology did not require such large amounts of capital investment. For example in Germany the average was about €150,000 and in the UK/ROI around £60,000. Large and small firms alike have invested in this technology and there was a tendency for firms that had gone beyond compliance to be the main investors in this type of equipment. The German sample showed that the investment in production technology led to a reduction in running costs.

6.5.7 Environmental Communication

Germany was the only country that participated to any great extent in environmental communication initiatives (Table 6.27). In Italy and the UK/ROI few firms were engaged in these activities. In Germany all the excellence firms and about 75% of the compliance plus firms took part in giving environmental information about the products sold and in encouraging guided visitor tours. Investment in these initiatives was quite small, averaging around €15,000.

Table 6.27. Environmental communication, number of firms, face-to-face interviews

Initiative	UK/ROI	Germany	Italy
Environmental report/statement	3	2	2
Environmental product information	6	24	8
Guided visitor tours	7	26	5
Hotline	–	1	1
Press conferences	1	7	1
Talks at meetings & conferences	5	14	5
Sponsorships	–	13	1
Sample size	33	34	30

In all the countries, these initiatives were predominately market driven and although the overall effect was increased cost the benefits in terms of improved market position, company image and competitiveness outweighed the additional costs.

6.5.8 Environmental Management Systems

The responses to the postal questionnaire showed that the single initiative that differentiates the average environmentally performing firms from the excellence firms was that relating to environmental management systems. Taking into account the fact that those firms with EMS will automatically have a higher than average number of initiatives, the difference between firms with EMS and those without is clear (Table 6.28).

Table 6.28. Average number of initiatives adopted by firms with and without an EMS, postal questionnaire

	Firms with EMS	Firms without EMS
	Average number of initiatives	Average number of initiatives
UK/ROI	6.65 (23)	2.77 (64)
Germany	7.95 (38)	3.77 (70)
Italy	4.08 (26)	1.43 (74)

Note: Figures in parenthesis denote number of firms.

In the UK/ROI and Germany there was a positive relationship between the firms that had an EMS and size (correlation coefficient of $r = 0.318$ and $r = 0.370$ both significant at $\alpha = 0.01$). In the UK/ROI there was evidence that larger firms (i.e. firms with more than 100 employees) were more likely to have an EMS whilst in Germany this employment threshold occurred at 50 employees. In Italy there was no relationship with firm size.

Table 6.29 shows, for the UK/ROI and Germany, the average number of eco-points (FTF interviews) for the environmental performance groups that did and did not implement certified environmental schemes such as ISO 14001. Firms that had achieved certification did undertake significantly more initiatives than those without, even after allowing for the additionality effect.

In Italy, none of the firms in the sample had implemented a certified scheme. However, all of the compliance plus and excellence firms had achieved some form of out-of-house certification during the 1990s. In addition, the 10 firms that were planning to implement an EMS within the next three years undertook, on average, more initiatives than firms that had no plans in this direction.

Table 6.29. Average number of initiatives adopted by firms with and without a Certified EMS by performance group, face-to-face interviews

	Firms with Certified EMS	Firms without Certified EMS
	Average number of eco-points	Average number of eco-points
UK/ROI		
Compliance only	11.00 (1)	7.50 (15)
Compliance plus	15.92 (4)	13.90 (13)
Germany		
Compliance only	–	8.40 (5)
Compliance plus	16.67 (1)	14.59 (17)
Excellence	22.53 (5)	20.50 (6)

Note: Figures in parenthesis denote number of firms.

However, the study was interested in not only establishing if a firm had a certified EMS system but whether the firm used management techniques thereby indicating an awareness of the issues and the need for quality management tools such as data collection and supplier evaluation. The results for each country are given in Tables

6.30 to 6.32. The data on environmental policy and improvement programme refer to whether or not the firm had a draft or official version in print.

Table 6.30. Environmental management tools by size group, UK/ROI, FTF interviews

Size	EP	IP	ET	DC	SE	Total firms
1-10	2	1	–	1	2	2
11-25	2	2	2	3	1	6
26-50	–	–	–	3	2	4
51-100	2	2	1	3	3	5
101-250	5	3	3	5	4	10
251-500	4	3	2	5	3	6
Total	15	11	8	20	15	33

EP Environmental policy, *IP* Improvement programme, *ET* Environmental team, *DC* Data collection, *SE* Supplier evaluation.

Table 6.31. Environmental management tools by size group, Germany, FTF interviews

Size	EP	IP	ET	RP	SE	Total firms
1-10	–	–	1	12	10	15
11-25	2	–	1	3	4	6
26-50	1	1	–	2	1	2
51-100	5	3	2	3	7	7
101-250	–	–	–	1	1	2
251-500	–	1	1	1	2	2
Total	8	5	5	22	25	34

EP Environmental policy, *IP* Improvement programme, *ET* Environmental team, *RP* Responsible person, *SE* Supplier evaluation.

Table 6.32. Environmental management tools by size group, Italy, FTF interviews

Size	EP	IP	ET	DC	SE	Total firms
1-10	–	–	–	–	–	–
11-25	5	5	4	6	–	6
26-50	1	1	1	2	3	3
51-100	2	5	8	7	9	10
101-250	7	8	8	8	7	11
251-500	–	–	–	–	–	–
Total	15	19	21	23	19	30

EP Environmental policy, *IP* Improvement programme, *ET* Environmental team, *DC* Data collection, *SE* Supplier evaluation.

In the UK/ROI and Italy there was a clear indication that the larger firms were more likely to make use of these environmental management tools. The results for Germany showed that there was no association between the size of the company and the ability to undertake management related activities. In terms of the performance groups, it was the compliance plus firms in the UK/ROI, the excellence firms in Germany and the beyond compliance firms in Italy that had an official version of the firms' environmental policy or environmental improvement program. Data collection for controlling environmental improvement was undertaken by 76% of the compliance plus firms in the UK/ROI and 47% of the compliance only firms. In Italy only 36% of the compliance only firms collected environmental data compared with all the compliance plus and excellence firms. There was no comparable data for Germany. In the UK/ROI just one compliance only firm had an environmental team compared with almost half of the compliance plus firms. Whilst in Germany it was only in the excellence performing group that environmental teams existed. In Italy just 18% of the compliance only firms had environmental team meetings compared with universal participation by the beyond compliance firms. The analysis also showed that in Germany evaluation of suppliers was undertaken to some extent by 60% of compliance only and compliance plus firms but by all the excellence firms. The Italian firms showed a similar pattern. In the case of the British/Irish firms 47% of compliance only firms evaluated the environmental performance of suppliers compared with 76% of compliance plus firms.

Limited data on the initial set-up costs of an EMS was available for Germany. It showed that the average cost of implementing an environmental management system was €18,000. In all countries the motivation for firms to undertake environmental management initiatives was the market. However, if the customer did not require EMS there was no incentive for firms to invest in it. As one British manager said "our customers do not demand that we have an EMS and it would not bring us any more business".

6.6 Drivers for Undertaking Initiatives

Tables 6.33 (PQ) and 6.34 (FTF) give the breakdown of the dominant drivers for each initiative along with the proportion of the firms, undertaking the initiative that cited each driver. The drivers considered were regulation, cost, market (customer and supplier demands) and health and safety (H&S). In some instances not all the firms surveyed responded to this question hence the data does not sum to 100 while in other cases firms gave more than one response thus the percentages may sum to over 100.

The tables show that with the exception of environmental management, the main drivers for the initiatives undertaken corresponded reasonable well between the two surveys. For some of the initiatives there was little variation between countries. In all countries, energy initiatives were clearly driven by cost considerations. Initiatives related to water use and treatment of waste water had the dual drivers of cost and regulation. Waste separation was driven by regulation in Italy and a combination of regulation and cost in the UK/ROI and Germany. However, the FTF results returned cost as the sole driver in Germany. Packaging initiatives were chiefly market driven except in the UK/ROI where cost was cited as an important driver (PQ results).

Table 6.33. Dominant drivers (cited by >20% of firms), postal questionnaire

Initiative	UK/ROI	Germany	Italy
Waste separation	Regulation 41% Cost 39%	Regulation 65% Cost 38%	Regulation 100%
Waste water treatment	Regulation 77%	Regulation 70% Cost 43%	Regulation 56%
Returnable packaging	Cost 32% Market 27%	Market 56% Cost 25%	Market 100%
Energy efficiency	Cost 82%	Cost 75%	Cost 88%
Water efficiency	Cost 76%	Cost 80%	–
Avoiding chemical ingredients	Market 72%	Market 54% Regulation 31%	Market 100%
Storage of hazardous substances	H&S 59% Regulation 38%	Regulation 71% H&S 44%	–
Regional sourcing	Market 68% Cost 34%	Market 51% Cost 37%	–
Organic raw materials	Market 100%	Market 67%	Market 100%
Environmental management	Regulation 48% H&S 30%	Regulation 42% Market 32%	H&S 62% Market 38%
Environmental report	Regulation 41% Market 41%	Regulation 50% Market 50%	–

Table 6.34. Dominant drivers (cited by >20% of firms), face-to-face interviews

Initiative	UK/ROI	Germany	Italy
Waste reduction	Regulation 45% Cost 26%	Cost 100% Regulation 50%	Regulation 59%
Packaging	Market 20%	Market 63%	Regulation 73%
Water use and protection	Regulation 50% Cost 36%	Cost 39% Regulation 23%	Regulation 61%
Energy	Cost 70%	Cost 57%	Cost 82%
Avoiding artificial ingredients	Market 65%	Market 47% Regulation 25%	Market 67%
Organic raw materials and regional sourcing	Market 95%	Market 57%	Market 70%
Environmental management	(Market 7%)	Market 67%	Market 59%
Environmental communications	Market 54%	Market 77%	Market 92%

Initiatives relating to organic raw materials and the avoidance of artificial ingredients, were almost exclusively market driven except in Germany where there was legislation banning the use of artificial substitutes in food products. The PQ data showed that initiatives relating to environmental management and environmental reporting (a component of environmental communications) were driven by regulation (including health and safety) and the market in roughly equal proportions. This did not agree with the findings of the FTF survey, which showed that the market was by far the principal driver.

Overall, in this industry, the role of the market and cost as motivators for implementing environmental initiatives was dominant with regulation playing a subordinate role.

6.7 Economic Impact of Initiatives

Respondents to the face-to-face interviews were asked to assess the implications of undertaking each of the environmental initiatives in terms of positive, negative or neutral effects on labour, sales, costs, competitiveness, profits, price, market position and image. At the aggregate level, using all the available data, there were 264 positive or negative effects arising from 206 initiatives in the UK/ROI, 297 from 188 initiatives in Germany and 190 from 188 initiatives in Italy. These gave a ratio of effects per initiative of 1.28, 1.58 and 1.01 respectively. Hence, Germany ranked in the prime position followed by the UK/ROI and lastly Italy.

Tables 6.35 to 6.37 give a breakdown by initiative for each of these effect variables for each country. In Germany and the UK/ROI the overall impact on the firm in terms of increased sales, reduced costs and associated increases in profits and competitiveness was extremely positive. In Italy these combined effects were also positive but only marginally. There was some variability in the effects between the countries for the same initiative. In the UK/ROI and Germany the cost driven energy efficiency initiative account for 15% and 13% of effects respectively whereas in Italy there was no recorded impact despite the fact that 73% of the firms were engaged in energy saving initiatives. It was the market driven initiatives of packaging in the UK/ROI, environmental management in Italy and environmental communications in Germany that recorded the highest number of effects. In all the countries, the switch to organic raw materials as a result of customer pressures did result in higher costs but these costs were more than offset by increased product price and sales, improved market position, competitiveness and company image. The tables show that in the UK/ROI there was little net gain in labour associated with the initiatives. However, in Germany there was a small net increase in labour whilst in Italy there was a substantial rise in labour. This increase was mostly associated with environmental management and the avoidance of artificial ingredients.

On the basis of these results there was evidence to support the hypothesis of a positive relationship between environmental performance and firm competitiveness.

Table 6.35. Economic impact, UK/ROI, face-to-face interviews

Initiative	Labour		Cost		Price		Sales		Position		Compe-titiveness		Profit		Image		Total
	↑	↓	↑	↓	↑	↓	↑	↓	↑	↓	↑	↓	↑	↓	↑	↓	
Waste reduction	2		2	5	1		2		2		5	1	8	3	8		39
Packaging	2	2	3	9	1	1	5		5		10		9	1	4		52
Water use and protection			1	8	3						4	1	8	1	2	3	31
Energy	1			7	1	1	1		2	2	11		10		2	1	39
Avoiding artificial ingredients			2	1		1	4		7		7		4		7		33
Organic raw materials and regional sourcing	1		9		8		7		5		7		3	1	5		46
Communication			4	2	1		2		3		5		4		3		24
Total	6	2	21	32	15	3	21		24	2	49	2	46	6	31	4	264

Table 6.36. Economic impact, Germany, face-to-face interviews

Initiative	Labour		Cost		Price		Sales		Position		Compe-titiveness		Profit		Image		Total
	↑	↓	↑	↓	↑	↓	↑	↓	↑	↓	↑	↓	↑	↓	↑	↓	
Waste reduction	3		3		5	1	1		1		2		5	1	11		33
Packaging	1	1	2	2	3		3	1	3	1	4	1	2	1	5		30
Water use and protection	1	1	6	7	4	2		1	1	1	1	1	7	2	7		42
Energy		1	4	8	3		1		3		2		7		9		38
Avoiding artificial ingredients				2		1			1		1				1		6
Organic raw materials and regional sourcing	5		8		4	2	8		9		7		4		10		57
Environmental management			1	1	2		1		1		3		3		1		13
Communication	5		4		4	1	15		14		11		7		17		78
Total	15	3	28	20	25	7	29	2	33	2	31	2	25	4	61		297

Table 6.37. Economic impact, Italy, face-to-face interviews

Initiative	Labour		Cost		Price		Sales		Position		Compe-titiveness		Profit		Image		Total
	↑	↓	↑	↓	↑	↓	↑	↓	↑	↓	↑	↓	↑	↓	↑	↓	
Waste reduction	3	2		2		2		2		2	2	4		3	2	7	31
Packaging																	0
Water use and protection	2		5	1	4												12
Energy																	0
Avoiding artificial ingredients	9		3		3		3		6		4				7		35
Organic raw materials and regional sourcing	6		6		5		2		7		4		1		7		38
Environmental management	10		6		7		9		13		10				13		68
Communication			1		1		1		1						2		6
Total	30	2	21	3	20	2	15	2	27	2	20	4	1	3	31	7	190

Table 6.38 shows the responses on the effects of the environmental initiatives from the SMEs that responded to the postal questionnaire. There was no comparable data for Italy hence it has been excluded from this analysis.

Table 6.38. Economic impact, postal questionnaire

Initiative	Sales		Labour		Costs		Profits	
	↑	↓	↑	↓	↑	↓	↑	↓
Waste separation								
UK/ROI	3		16	1	19	13	13	9
Germany	7	6	43	3	33	26	18	21
Waste water treatment								
UK/ROI			2		12	3	2	6
Germany	1	3	15	2	17	12	8	16
Returnable packaging								
UK/ROI	6		5	2	9	6	3	4
Germany	15	2	18	4	21	7	11	14
Energy efficiency								
UK/ROI				1	3	9	10	1
Germany	4	2	8	1	9	23	17	3
Water efficiency								
UK/ROI	1		1		3	8	8	
Germany	1	3	8		6	26	18	7
Avoiding artificial ings.								
UK/ROI	23		3	2	9	8	17	4
Germany	13	2	7	3	11	4	8	4
Storage haz. substances								
UK/ROI			4	1	14	1	1	5
Germany	1	5	21	1	29	2	1	12
Regional sourcing								
UK/ROI	15		5	3	8	14	5	3
Germany	27	2	18	4	12	20	26	7
Organic products								
UK/ROI	21		8	2	16	1	14	2
Germany	17	1	10		16		12	4
Env. management								
UK/ROI	5		5		13	1	1	5
Germany	6	3	18		22	2	2	14
Env. report								
UK/ROI	2		5		7	1	2	
Germany	3	1	9		11			4
Total Impact								
UK/ROI	76		54	12	113	65	76	39
Germany	95	30	175	18	187	122	121	106

In the UK/ROI there were 435 total effects from 330 initiatives giving a ratio of 1.32 effects per initiative. The analogous figures for Germany were 854 effects from 566 initiatives, a ratio of 1.51 effects per initiative. There was some variability in effects between countries for the same initiative. For example, for the organic raw materials initiative 68% of the British/Irish firms recorded an increase in sales whilst in Germany only 50% of firms undertaking this initiative experienced an increase in sales.

The experience of the firms that were interviewed was reasonably positive in terms of falling costs, increasing sales and profits and although this position was endorsed by the results from the postal survey, in the case of Germany the overall net gain was marginal. The effect on the firm of environmental management initiatives was negative in the UK/ROI and Germany with firms indicating increased costs, little improvement in sales and hence falling profits. The additional costs were most likely related to the extra labour required to implement the initiative. The total impact on labour was substantially greater in Germany than in the UK/ROI. In the UK/ROI waste separation was the biggest single contributor to this overall net gain in labour. In Germany the three initiatives of waste separation, storage of hazardous substances and environmental management were responsible for just over half of this net gain in labour.

One focus of the research was examining the relationship between the environmental performance of the firm, when ranked as being compliance only, compliance plus and excellence, and the economic impact that undertaking these environmental initiatives had on the firm. Constraints within the datasets meant that this analysis could be carried out only with data from the postal survey and also excluded Italy because of the incompleteness of the Italian data. When the responses from the PQ for the UK/ROI and Germany are combined, the distribution of economic effects, the total number of initiatives undertaken and the effects to initiative ratio for each environmental performance group is given in Table 6.39.

The final column in Table 6.39 shows that the ratio of effects to initiatives remains unchanged as firms progress from compliance only to compliance plus. However, when firms achieve excellence status (i.e. undertakes 6 or more initiatives) there was a corresponding increase in the effects to initiatives ratio. A substantial part of this increase was due to the additional labour necessary to support management and increased participation in waste handling (waste water treatment, waste separation, returnable packaging) initiatives. From the firm's perspective a positive economic effect would be increased sales, reduced costs and consequently raised profits. Reworking the ratio of effects to initiatives to include only these factors shows that the ratio for compliance only firms was 0.46; for compliance plus the ratio was 0.64 and 0.62 for excellence firms. This would indicate that compliance only firms were not capable of achieving the same economic benefits as the better environmental performing firms and that there were few purely economic benefits to firms in undertaking more initiatives.

Table 6.39. Economic impact, performance groups and size, UK/ROI and Germany, PQ

	Sales		Labour		Costs		Profits		TE	TI	E/I
	↑	↓	↑	↓	↑	↓	↑	↓			
C only	13	2	14	7	14	8	2	10	70	50	1.40
C plus	47	9	58	9	72	57	56	35	343	245	1.40
Ex.	111	19	157	14	214	122	139	100	876	601	1.46
1-25	111	21	126	23	141	99	110	74	705	478	1.47
26-50	26	1	30	2	46	26	23	28	182	139	1.31
51-100	15	3	31	5	58	20	30	24	186	127	1.46
> 100	19	5	35	–	52	40	43	19	213	145	1.47
Total	171	30	222	30	297	185	206	145	1286	889	1.45

C only Compliance only, *C plus* Compliance plus, *Ex.* Excellence, *TE* Total effects, *TI* Total initiatives, *E/I* Effects per initiatives.

In terms of a relationship between the size of the firm and the economic impact of environmental initiatives Table 6.39 shows that at the aggregate level it was the two extremes of the distribution that had the highest ratio. This remains the case when the ratios for the positive economic benefits only were considered. Small firms (less than 25) and large firms (over 100) appeared to experience more economic benefits per initiative than the medium sized firms.

6.8 Obstacles to Adoption of Environmental Initiatives

At the face-to-face interviews the firm managers were asked to rank in order of importance the factors that made it difficult to introduce environmental initiatives into the organisation. Table 6.40 shows the ranking by country for each of these factors. The ranking is based on the frequency distribution of answers with a “1” indicating the most frequently cited problem, a “2” the next most important factor and so on.

In terms of the constraints to implementing environmental initiatives there was a general agreement across all the countries that the main limitations were the difficulty faced in raising capital for a new investment and the length of the associated payback period. Shortage of the correct skills and expertise was an issue in Germany but was of little or no concern in Italy or the UK/ROI. British/Irish firm managers considered that from a management point of view there were too many other priorities and hence environmental issues came low down the list of concerns. In addition, the British/Irish firms believed that the shortage of management time to address such issues was an important constraint. Management concerns were not considered to be serious obstacles in either Italy or Germany. The uncertainty of regulation in planning for technology changes was important in Italy and

Germany as was the related issue concerning regulation and the lack of encouragement to firms to adopt clean technology. These obstacles that were allied to regulation were not important in the UK/ROI. Quality and cost of advice from external sources were not thought to be significant in any of the countries.

Table 6.40. Ranking of obstacles to adoption, face-to-face interviews

Problem	UK/ROI	Germany	Italy
Difficult to raise required capital	1	1	2
Payback period on clean technology is too long	4	2	1
Shortage of required skills and expertise	7	4	–
Clean technology is risky and unproven	10	4	–
Management priorities are focused on other issues	2	7	6
Lack of management time	3	6	6
Profit motive supersedes environmental initiatives	5	6	5
Lack of good quality advice	6	8	8
Lack of help from suppliers	7	8	8
Environmental consultancy services are expensive	8	6	7
Regulations are too uncertain to plan for new technology	10	3	3
Regulation does not encourage clean technology solutions	9	5	4

A breakdown of these obstacles by environmental performance group is given in Table 6.41. With the exception of the profit motive, there was little to distinguish between the two performance groups in the UK/ROI. Both groups were concerned with finding capital for investment, payback and management issues. The compliance plus firms were more anxious about the need for the firm to make a profit irrespective of other concerns. With the exception of the problem associated with the cost of consultancy services, the German responses by performance group showed little difference between the groups. Compliance only firms were more concerned than firms in the other performance groups about the cost of obtaining advice. Italian firms showed little variability between the groups except for management priorities where compliance only firms considered this factor a severe restriction

There was no association between the size of the firm and the ranking of the obstacles, small and large firms alike endorsed the general pattern.

Table 6.41. Ranking of obstacles to adoption by environmental performance groups, face-to-face interviews

Problem	UK/ROI		Germany			Italy		
	C	C+	C	C+	E	C	C+	E
Difficult to raise required capital	1	2	1	1	1	2	1	1
Payback period on clean technology is too long	4	2	2	2	2	1	2	1
Shortage of required skills and expertise	7	6	1	2	–	–	–	–
Clean technology is risky and unproven	8	–	1	4	3	–	–	–
Management priorities are focused on other issues	3	1	–	6	4	2	5	–
Lack of management time	2	4	–	6	3	3	–	–
Profit motive supersedes environmental initiatives	7	3	2	5	–	3	5	–
Lack of good quality advice	5	6	–	–	4	–	5	–
Lack of help from suppliers	7	5	–	7	–	6	–	–
Environmental consultancy services are expensive	6	7	2	6	4	6	5	–
Regulations are too uncertain to plan for new technology	–	7	1	3	3	4	3	–
Regulation does not encourage clean technology solutions	8	7	–	5	3	5	4	2

C Compliance only, *C+* Compliance plus, *E* Excellence.

6.9 Environmental Strategy

Company strategy as it relates to environmental issues was thought to be influential in the decision by a company on the implementation of clean technology. On the basis of the postal survey results (Table 6.42, note that some firms indicated two main strategies) there was variation between the countries in the strategies that were followed.

In the British/Irish sample firms, regulation was the preferred strategy of about half the companies surveyed followed by eco-practice and eco-products. Cost strategies accounted for about 20% of responses. In an industry where the environmental initiatives were driven by a combination of regulation, market and cost this was to be expected. German companies adopted regulation, cost and eco-practice strategies in almost equal proportions. As in the UK/ROI this was a reflection of the initiative drivers (see Table 6.33). In Italy firms adopted either a regulation or eco-practice strategy and it was unexpected that cost as a strategy did not feature more notably considering that almost 40% of the firms adopted cost driven energy saving initiatives.

Table 6.42. Firm environmental strategy, number and percentage (in brackets) of firms, postal questionnaire

Main focus	Compliance only	Compliance plus	Excellence	All firms
UK/ROI				
Regulation	16 (53.3)	18 (54.5)	9 (37.5)	43 (49.4)
Cost savings	4 (13.3)	9 (27.3)	4 (16.7)	17 (19.5)
Promoting brand by eco-practice	10 (33.3)	14 (42.4)	10 (41.7)	34 (39.1)
New markets for eco-products	–	2 (6.1)	3 (12.5)	5 (5.7)
Promoting eco-products	4 (13.3)	13 (39.4)	6 (25.0)	23 (26.4)
Total firms	30	33	24	87
Germany				
Regulation	11 (42.3)	9 (33.3)	20 (36.4)	40 (37.0)
Cost savings	7 (26.9)	12 (44.4)	21 (38.1)	40 (37.0)
Promoting brand by eco-practice	6 (23.1)	9 (33.3)	27 (49.1)	42 (38.9)
New markets for eco-products	2 (7.7)	–	12 (21.8)	14 (13.0)
Promoting eco-products	3 (11.5)	4 (14.8)	19 (34.5)	26 (24.1)
Total firms	26	27	55	108
Italy				
Regulation	36 (47.4)	3 (13.6)	–	39 (39.0)
Cost savings	6 (7.9)	1 (4.5)	–	7 (7.0)
Promoting brand by eco-practice	17 (22.4)	11 (50.0)	2 (100.0)	30 (30.0)
New markets for eco-products	3 (3.9)	–	–	3 (3.0)
Promoting eco-products	14 (18.4)	7 (31.8)	–	21 (21.0)
Total firms	76	22	2	100

In terms of difference between the environmental performance groups, in Germany there was a significant difference across groups in the adoption of strategies to promote brand by eco-practice or develop new markets for eco-products with the excellence firms preferring these strategies. In Italy, compliance only firms favoured regulation whilst compliance plus and excellence firms showed a distinct preference for adopting good environmental practice as the primary strategy. In the UK/ROI there was no significant difference in the strategy adopted across performance groups.

Table 6.43 gives the breakdown of answers to a complementary question that asked about the factor that had the most influence on the final decision as regards environmental issues. In Germany and the UK/ROI, regulation was not as important as had been the case when overall company strategy was examined whereas in

Italy regulation as a strategy retained its key position. Management awareness was the most important factor in the UK/ROI and supports the view among the British/Irish firms that management priorities and time were constraints in introducing clean technology. In Germany management awareness and customer demands were equally influential. In a sector where the market was the main driver for many of the initiatives it was expected that customer demands would have a strong influence on the decision to implement environmental initiatives. When these data were tested for differences between environmental performance groups a significant difference (at $\alpha = 0.01$) was found in the Italian figures reinforcing the importance of regulation amongst the compliance only firms. In terms of firm size, the British/Irish firms showed management awareness was more important in small firms (less than 25 employees) than in the larger companies. In Italy and Germany there was no relationship between firm size and the factor that had the most influence in the final decision making process.

Table 6.43. Influences on final decision regarding environmental issues, number and percentage (in brackets) of firms, postal questionnaire

Main influence	Compliance only	Compliance plus	Excellence	All firms
UK/ROI				
Regulation	3 (10.7)	12 (36.4)	6 (27.3)	21 (25.3)
Management awareness	9 (32.1)	12 (36.4)	8 (36.4)	29 (34.9)
Cost	8 (28.6)	3 (9.0)	3 (13.6)	14 (16.9)
Customer demands	8 (28.6)	6 (18.2)	5 (22.7)	19 (22.9)
Total firms	28	33	22	83
Germany				
Regulation	5 (26.4)	5 (20.8)	7 (15.9)	17 (19.5)
Management awareness	4 (21.0)	7 (29.2)	17 (38.6)	28 (32.2)
Cost	2 (10.5)	5 (20.8)	6 (13.6)	13 (14.9)
Customer demands	8 (42.1)	7 (29.2)	14 (31.8)	29 (33.3)
Total firms	19	24	44	87
Italy				
Regulation	38 (50.0)	2 (9.1)	–	40 (40.0)
Management awareness	8 (10.5)	11 (50.0)	1 (50.0)	20 (20.0)
Cost	4 (5.3)	–	–	4 (4.0)
Customer demands	26 (34.2)	9 (40.9)	1 (50.0)	36 (36.0)
Total firms	76	22	2	100

In Sect. 6.7 there was evidence to show that there were economic benefits to be achieved from undertaking environmental initiatives. Table 6.44 shows that in the UK/ROI almost half the firms supported the view that the main impact of environmental legislation was improved firm productivity. This view was held by about one-third of the companies in both Italy and Germany. Only in Germany did a substantial proportion of firms consider that environmental initiatives would reduce overall firm competitiveness. The only country where there was a significant difference (at $\alpha = 0.01$) between the environmental performance groups was Germany with a high proportion of excellence firms supporting the improved competitiveness view. In addition, in Germany, there was a difference in the views expressed with larger firms noting the productivity benefits and the smaller firms supporting the opposite position. There was no relationship between firm size and this variable in Italy or the UK/ROI.

Table 6.44. Perception of the main impact of environmental legislation, number and percentage (in brackets) of firms, postal questionnaire

	Compliance only	Compliance plus	Excellence	All firms
UK/ROI				
Reduced competitiveness	4 (14.8)	5 (15.2)	4 (17.4)	13 (15.7)
Improvement for the good of society	10 (37.0)	13 (39.4)	10 (43.5)	33 (39.7)
Improved productivity	13 (48.2)	15 (45.4)	9 (39.1)	37 (44.6)
Total firms	27	33	23	83
Germany				
Reduced competitiveness	8 (44.4)	11 (50.0)	14 (27.5)	33 (36.3)
Improvement for the good of society	5 (27.8)	8 (36.4)	15 (29.4)	28 (30.7)
Improved productivity	5 (27.8)	3 (13.6)	22 (43.1)	30 (33.0)
Total firms	18	22	51	91
Italy				
Reduced competitiveness	10 (13.5)	2 (9.1)	–	12 (12.2)
Improvement for the good of society	35 (47.3)	18 (81.8)	2 (100.0)	55 (56.1)
Improved productivity	29 (39.2)	2 (9.1)	–	31 (31.6)
Total firms	74	22	2	98

Table 6.45 shows the responses from the FTF interviews regarding the environmental strategy adopted. In all the countries the results differed to a certain extent from the findings of the PQ. Cost saving as an environmental strategy became more important in the UK/ROI, especially amongst the compliance only firms, and Italy whilst the influence of regulation diminished. Except in Italy, the emphasis on environmentally proactive strategies, particularly eco-products was somewhat diminished. The German results showed that compliance only firms did not follow a mix of strategies as shown in the PQ but were only involved in regulation or cost savings strategies. There was evidence that when German firms move beyond compliance they adopted environmental strategies that were proactive in terms of improving the firm's market position by differentiating themselves from their competitors through eco-practices or eco-products. However, these differ-

ences between compliance and beyond compliance firms were not statistically significant.

Table 6.45. Environmental strategy, number and percentage (in brackets) of firms, FTF

	Compliance only	Compliance plus	Excellence	All firms
UK/ROI				
Regulation	5 (33.3)	5 (31.3)		10 (32.3)
Cost saving	8 (53.3)	6 (37.5)		14 (45.2)
Eco-practices	2 (13.3)	3 (18.8)		5 (16.1)
Eco-products	–	2 (12.5)		2 (6.5)
Total firms	15	16		31
Germany				
Regulation	3 (75.0)	3 (20.0)	1 (9.1)	7 (23.3)
Cost saving	1 (25.0)	5 (33.3)	3 (27.3)	9 (30.0)
Eco-practices	–	–	2 (18.2)	2 (6.7)
Eco-products	–	7 (46.7)	3 (27.3)	10 (33.3)
New markets for eco-products	–	–	2 (18.2)	2 (6.7)
Total firms	4	15	11	30
Italy				
Regulation	6 (60.0)	1 (6.2)	–	7 (24.1)
Cost saving	2 (20.0)	4 (25.0)	1 (33.3)	7 (24.1)
Eco-practices	–	–	–	–
Eco-products	–	2 (12.5)	1 (33.3)	3 (10.3)
New markets for eco-products	2 (20.0 0	9 (56.3)	1 (33.3)	12 (41.4)
Total firms	10	16	3	29

6.10 Competitive Strategy

At the FTF interviews managers were asked about the competitive strategy of the company. Firms were categorised as belonging to one of the following groups depending on the firm's competitive strategy: cost leadership with an emphasis on low cost, differentiation where the emphasis was on uniqueness, or niche products with a focus on a particular segment of the market. Table 6.46 shows the resulting distribution, which exhibited considerable variation across countries.

Table 6.46. Competitive strategy adopted by environmental performance group, number and percentage (in brackets) of firms, face-to-face interviews

	Cost leadership	Differentiation	Niche	All firms
UK/ROI				
Compliance only	5 (31.3)	5 (31.1)	6 (37.5)	16 (100.0)
Compliance plus	10 (58.8)	6 (35.3)	1 (5.9)	17 (100.0)
All firms	15 (45.5)	11 (33.3)	7 (21.2)	33 (100.0)
Germany				
Compliance only	–	1 (20.0)	4 (80.0)	5 (100.0)
Compliance plus	2 (11.1)	5 (27.8)	11 (61.1)	18 (100.0)
Excellence	2 (18.2)	5 (45.4)	4 (36.4)	11 (100.0)
All firms	4 (11.8)	11 (32.4)	19 (55.9)	34 (100.0)
Italy				
Compliance only	3 (27.3)	5 (45.4)	3 (27.3)	11 (100.0)
Compliance plus	2 (12.5)	8 (50.0)	6 (37.5)	16 (100.0)
Excellence	–	2 (66.7)	1 (33.3)	3 (100.0)
All firms	5 (16.7)	15 (50.0)	10 (33.3)	30 (100.0)

In the UK/ROI almost half of the firms followed a cost leadership strategy with about one-third adopting a differentiation strategy and the remainder a niche strategy. In Italy about half of the firms had a differentiation strategy, one-third niche and the remaining one-sixth a cost strategy. In Germany the situation was again different with niche strategies dominating (56% of firms) followed by differentiation (32%) and less than 12% of firms pursuing a low cost competitive strategy. *A priori*, it was anticipated that the better environmental performing firms would be clustered in the differentiation or niche categories but this was not the case for the UK/ROI where proportionately more compliance plus firms adopted a low cost strategy compared with compliance only firms. The differences between these two performance groups was statistically significant at $\alpha = 0.05$. In Italy and Germany there was no difference between the environmental performance groups regarding the type of strategy adopted. Also the expectation that the firms that pursued a cost leadership strategy would undertake fewer initiatives than those with a differentiation or niche strategy was only evident in Italy (Table 6.47).

Table 6.47. Competitive strategy and average number of eco-points, face-to-face interviews

	UK/ROI	Germany	Italy
Cost leadership	12.04	16.75	10.78
Differentiation	11.52	17.33	13.09
Niche	8.67	14.98	12.78

When evaluating the relationship between the competitive strategies and the environmental strategies adopted by the firms there was an assumption that firms with a differentiation or niche competitive strategy would have had an environmental strategy that reflected this approach i.e. they would have implemented proactive environmental strategies such as environmentally friendly practices and products. The data in Table 6.48 shows that this was not the case in the UK/ROI. There was evidence in Italy and Germany that firms with a differentiation strategy did implement proactive environmental strategies. However, this was not so with firms that had focused competitive strategies where regulation and efficiency strategies were undertaken by at least half of these firms.

Table 6.48. Competitive and environmental strategy adopted, number of firms, FTF

	Cost leadership	Differentiation	Niche	Total
UK/ROI				
Regulation	4	5	1	10
Eco-efficiency	6	5	3	14
Eco-practices	3	–	2	5
Eco-products	1	1	–	2
Eco-markets	–	–	–	–
Total firms	14	11	6	31
Germany				
Regulation	–	3	4	7
Eco-efficiency	2	–	7	9
Eco-practices	1	1	–	2
Eco-products	1	5	4	10
Eco-markets	–	–	2	2
Total firms	4	9	17	30
Italy				
Regulation	2	2	3	7
Eco-efficiency	2	3	2	7
Eco-practices	–	–	–	–
Eco-products	–	3	–	3
Eco-markets	1	6	5	12
Total firms	5	14	10	29

6.11 Summary

In terms of environmental performance, as measured by the average number of initiatives undertaken per firm, Germany was the top performer. There was some evidence from the German and Italian samples to support the hypothesis that the more competitive firms undertake more environmental initiatives. Firms that participated in R&D and also served an export market were more likely to have better environmental performance irrespective of size.

The two surveys showed some differences in the type of initiative undertaken. On the basis of the postal survey, there was a distinction between countries in the type of initiative that compliance only firms undertook. In the British/Irish sample the focus was on market driven initiatives whereas in the German sample, compliance only firms were engaged in mostly waste and packaging initiatives, with limited participation in market initiatives whilst in Italy compliance only firms concentrated on waste and energy efficiency initiatives. There was a lack of participation in environmental management initiatives irrespective of nationality. Compliance plus firms in all the countries were engaged in a more extensive range of initiatives but to a much lesser degree than the excellence performing firms, particularly as regards management initiatives, water protection and use, and market driven initiatives. The results from the face-to-face interviews showed that compliance only firms were not especially involved in market driven initiatives. There was a high degree of adoption of most initiatives irrespective of performance group and the initiatives that most clearly differentiated the compliance plus firms from the compliance only firms were environmental communication, water use and protection, and market initiatives.

Overall, the role of the market and cost as motivators for implementing environmental initiatives was dominant with regulation playing a more limited part. The aggregate economic impact of the initiatives was positive and there was evidence that environmental performance was associated with improved economic performance.

There was some variation between the countries in the environmental strategies followed, which, on the whole did reflect the environmental drivers. Regulation strategy was important along with market strategies aimed at improving both practices and products. Cost strategies were important in Germany and to a much lesser degree in the UK/ROI and Italy. There was evidence in Italy and Germany that firms with a differentiation strategy did implement proactive environmental strategies.

7 Sources of Advice (SOA)

7.1 Introduction

The importance of information and advisory sources in facilitating the take up of clean technologies is the third key area of interest in this research. This aspect centres on a need for knowledge and understanding and independent advice (see Sect.1.2). Dissemination of information and advice depends on a number of factors including industry structure, with firms in fragmented industries remaining relatively poorly informed. Hence, trade associations have an important role to play in providing independent advice. Information and advisory sources include universities, R&D organisations, official government sources, business-to-business networks, consultants, along with the help and advice provided by suppliers and customers.

The focus of enquiry in this research was on the advisory process, the drivers and obstacles to seeking advice, the impact of the advisory services and from the viewpoint of the manufacturing firm, those advisory services they considered most useful. Various categories of advisory service such as suppliers, customers, universities, research associations, professional associations, public bodies and private consultancies are considered.

7.2 Sampling Procedure

In total, 320 providers of advice from the four countries participated in face-to-face interviews (see Table 7.1 and Figs. 7.1 and 7.2 for more details). A questionnaire that focused upon clients' demands, the type of professional services supplied, the effectiveness of advisory services, major obstacles faced by firms in implementing environmental initiatives and so on was used in the interviews. Interviewees targeted included regulators, private consultants, members of business networks, public bodies, industrial associations, customers and suppliers of machinery and materials. In addition, users (i.e. SMEs) were asked who they considered to be their most important advisors for environmental legislation and initiatives.

Table 7.1 shows the number of advisory services interviewed by country and industrial sector. General advisors were those advising across a range of sectors. In the UK/ROI, a broad cross-section of advisors was interviewed (Fig. 7.1). A third were private consultants, 16% were public bodies and a significant number of non-government organisations (NGOs), government bodies and universities and research organisations were also contacted. In Germany, suppliers of machinery and materials were an important category of interviewees (Fig. 7.2).

In Italy, interviews were carried out at a professional environmental fair in Ricicla. The majority of those interviewed were from private consultancy companies (83%) with the remainder from non-profit organisations (government or employers' associations), which provided environmental information and support.

Table 7.1. Number of advisory sources surveyed by sector

	Furniture	Textile & finishing	Fruit & vegetable processing	General[a]	Total
UK/ROI	22	18	22	38	100
Germany	36	44	21	11	112
Italy	24	32	52	–	108

[a] Not sector specific advice.

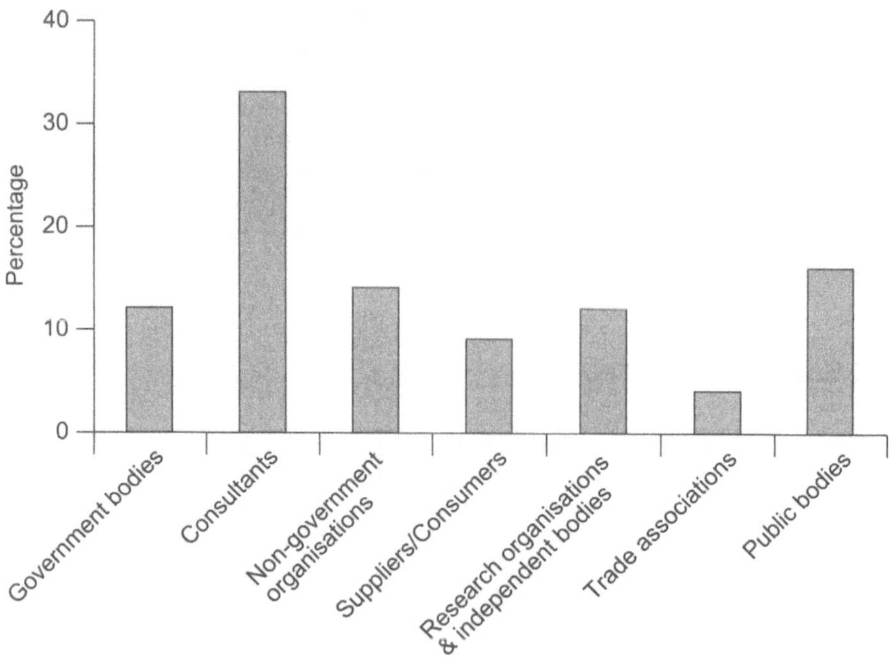

Fig. 7.1. Advisors interviewed in the UK/ROI

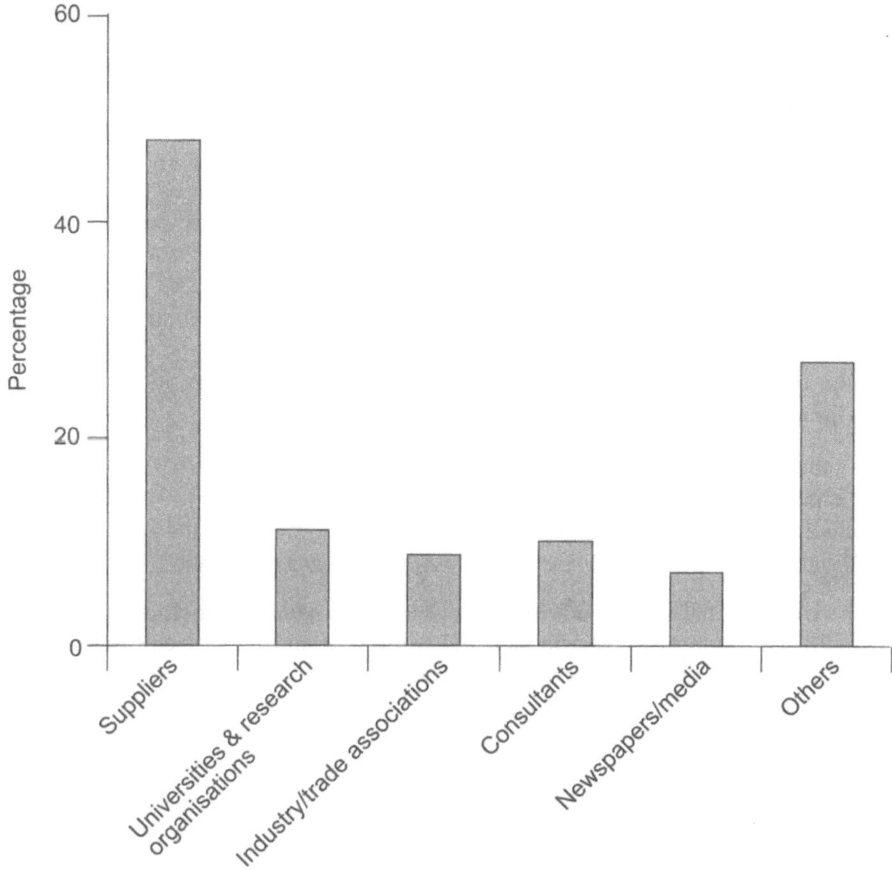

Fig. 7.2. Advisors interviewed in Germany

7.3 Sample Profile

In the UK/ROI, a wide variety of bodies offering advisory services were contacted. More than half of these had been established in the 1990s and they ranged in size from 1–300 employees with an average staff size of 16. Services offered included specialist technical advice, environmental audits, management consultancy and training. Many firms were reluctant to reveal their consultancy rates. However, in a number of cases a rate of €550/day was charged; in other cases clients were charged a 50% fee and sometimes the service or a component of the service was free.

In Germany, most of the agencies interviewed were also established in the 1990s and specialised in environmental and safety consulting. Three interviewees reported quality management and general management system consulting to be their main field of business, focusing more on management systems than on technical questions. The daily consultancy rate varied from €500 to €1200 with an average rate of €750.

The average number of employees in the Italian environmental consultancies interviewed was 13, with some variation depending on location. Consultancy firms situated in the northern regions tended to be large (over 10 employees on average) while companies in the South and Central regions relied on a network of external independent collaborators and technical professionals. The reason for this organisational difference was due to more limited and less varied opportunities. The average age of the consultancy firms was 7 years with about one-third of firms more than 10 years old.

The types of advice provided by the organisations that were interviewed fell into three broad categories:

- Technology specific (e.g. redesigning production processes, new products, implementing and monitoring procedures).
- Environment specific (e.g. waste treatment, control of water and air emissions).
- Organisation specific (advice covering the whole activity of the firm).

In the UK/ROI, 63% of the advisors interviewed provided technology specific advice, 28% environment specific, and 9% offered organisation specific advice. In Italy the type of advisory service was technology specific in 41% of cases, and environment specific and organisation specific in 36% and 23% respectively. In Germany, the majority were technology specific.

7.4 Advice Delivery

The advisors acknowledged that support needs among SMEs varied enormously; thus every enquiry was dealt with on an individual basis. The most common support services included free or subsidised briefings, seminars, workshops targeting specific business activities or environmental requirements, guiding SMEs through tailored and affordable environmental programmes, training staff and providing the support necessary to implement programmes.

7.4.1 Mode of contact

Advisory services in the UK/ROI established contacts with firms mainly through referrals (49%) particularly from government agencies. Other modes of contact included advertisements, mailshots, personal visits, telemarketing and seminars. In Germany, contact with clients was made by means of active advertisement through trade fairs, the Internet and yellow pages. While in Italy personal contacts (54%) predominated in the North West and South. Less personalised forms of advertising, such as participation in public events, yellow pages and Internet advertising were more common in the North East.

7.4.2 Number of contacts

Information about the activity of environmental consultants can provide some indication about the characteristics of the clients, their demands for technical assistance and their specific production problems in implementing environmental measures. The number of clients per year, as an example, is considered under specific conditions as an indicator of the firms' need for advice on environmental related issues (OECD 1999).

The advisors interviewed in the UK/ROI on average, catered for about 75–100 clients per year, which included small, medium as well as large firms. The frequency of contacts depended largely on the nature of the problem and the willingness of the firm to maintain further contact with the advisory bodies. In many cases (especially among small SMEs) the clients were not keen on long-term contacts because of the cost and time involved.

In Germany, 30% of the total consultancy firms interviewed had fewer than 100 clients, 20% offered their services to about 100–500 clients, 40% had more than 500 clients and 10% were reluctant to reveal the numbers. However, the number of clients does not represent the amount of counselling, since there are standardised advisory processes like waste separation, which would not require additional counselling. On the other hand, working on intensive consulting projects like implementing an EMS in an international group would require a proportionately greater amount of counselling. German experience showed that the number of meetings necessary differed according to the clients' needs and therefore could not be quantified exactly. On the whole, a shorter period of time is expected, if a customer relationship has already been established and if a high degree of open-mindedness can be found on the part of the SME.

In Italy, 32% of advisory services dealt with less than 50 clients per year, 41% handled between 51 and 300 clients with 27% working with more than 300 clients. During the interview, some advisors said that they keep in contact with former clients and consider the 'small advice' given frequently by phone (for example about how to fill in forms for the local Environment Protection Agency) as a relevant component of their professional tasks. As before there was a regional divide with advisors located in North West regions having a larger number of clients than those in the Central and Southern regions.

7.4.3 Problem identification and nature of involvement

In most cases SMEs contacted the sources of advice with specific enquiries, in other instances the problem was identified by the advisors. In Italy, in about 60% of the cases, the firm and the consultant identified the problem by working together. Once the problem was identified, the duration of time taken to provide an effective solution depended on the intensity of the problem. In the UK/ROI and Germany the time between the first contact and a decision on appropriate action varied extensively between a few weeks and a year. In the case of Italy the length of time involved in identifying and providing a solution to a problem varied between 4–5 months.

The solutions most frequently recommended by consultants involved changes in production processes; technical monitoring of production; technical answers to product mismanagement; assistance in innovation processes and improving the quality of products. Almost all the respondents maintained that, in order to improve their environmental performance, SMEs needed to reorganise production processes and adopt new machinery, compatible with higher environmental standards. The nature of the relationship between firms and the sources of advice was usually intensive, with the exception of the German suppliers of materials, who felt that intensive advice to customers was not important.

7.5 Implementation of Advice

7.5.1 Drivers for seeking advice

According to the advisors, the major driving force behind firms seeking advice in the UK/ROI furniture sector is cost. In the textile and food sector, regulation and cost featured as important drivers. In Germany the main driver is regulation in the furniture and food sector and market requirements in the textile sector while in Italy it is a combination of regulation and market with no difference between sectors.

Cost, as a major driver in the furniture sector in the UK/ROI is a reflection of the importance of cost in this sector. Firms in this sector generally adopt the suppliers' equipment for cost savings, with the EPA compliance coming as a *bonus*. Some sources of advice, observed that they found some firms to be resistant to regulations. One advisory source cited an example where they contacted 1,000 customers and offered a free compliance audit, they did not receive a single reply. As strict regulations apply to VOC emissions and toxic wastes associated with varnishes and glues in the German furniture sector, regulation as a driver was to be expected.

In addition, SOA cited the following as reasons why SMEs seek advice.

- Specific environmental legislation or regulatory matters.
- Supply chain pressure.
- Customers demanding proof of environmental management performance.
- Financiers or insurers seeking assurances regarding issues of environmental risk.
- Specific environmental concerns or problems.

7.5.2 Obstacles in the take up of advice (as identified by SOA)

UK/ROI

In the UK/ROI, lack of time and resources were identified by the SOAs as constraining the take up of advice by SMEs. The advisors were of the opinion that SMEs in the furniture sector lacked awareness. Even if firms were aware, they often failed to give the issue top management commitment, preferring to include the responsibility within existing quality management. Also, SMEs have a strong production focus, there are few resources available for environmental projects and firms are very unwilling to invest without a payback (maximum 2-year payback). Moreover many firms are slow to use training, and this has implications for the adoption of new technology/approaches.

The ease of access to landfill was thought to be an important obstacle in the adoption of environmental initiatives, as firms are not facing financial pressure to move to reduction and recycling. Local authorities are perceived as not enforcing waste regulations leading to a situation where there is widespread non-compliance, either through ignorance or on purpose.

Typical environmental problems associated with the UK/ROI textile industry involved effluent, waste and energy usage. SMEs in the textile sector often lacked sophisticated management systems and were more rigid and focused on traditional methods. In general SMEs concentrate on the core production activity and therefore expenditure on peripheral issues is met with reluctance.

The main problems for the UK/ROI food industry were out-of-date technology/machinery, unwillingness to change, lack of education and lack of awareness. In addition, SMEs in this sector do not feel that environmental issues were of any great concern to their business and therefore were unwilling to put the resources into such activities. The management of environmental issues was usually undertaken by the health and safety manager, production manager or quality manager who did not have the time to carry out the extra environmental work. These problems were characterised to a greater extent in very small firms (1–10 employees).

Germany

Lack of capital required to incorporate changes following advice was considered to be a major problem in all three sectors in Germany. Other typical problems identified in the furniture sector were the disinterested attitude of management to change. This unwillingness to change, stems from the fact that most of these firms anticipated increased costs following an advisory input. Lack of resources and time to deal with these issues was an additional constraint.

In the textile sector, the resistance to adopting advice on clean technology was related to cost, the existing corporate culture and the unsatisfactory qualifications of personnel in the firm.

Difficulties concerning finance, which are often decisive in the case of small companies, figured as prominent obstacles in the German food sector. Furthermore medium-sized firms were unwilling to take the risk of an investment because the profit margins are small. Regarding typical technical, material or product problems, 33% of the consultants stressed that material input (use of certain chemicals that create wastes such as chlorinated hydrocarbons) was a problem.

Italy

In Italy, the cost of the advisory service was identified as a major problem by 48% of the respondents. There was an innate apprehension that a long-term relationship with an advisor would lead to excessive costs. The second most cited constraint, was the reluctance on the part of SMEs to change, mainly due to the fact that most SMEs prefer to seek short-term solutions to problems. The lack of general technical knowledge, trust and quality of information were cited as constraints on consultants' action. These obstacles were present in SMEs irrespective of industry sector.

7.5.3 Typical technological & organisational obstacles

SOAs interviewed were asked what in their opinion were the major technological and organisational problems faced by the SMEs in the adoption of environmental initiatives.

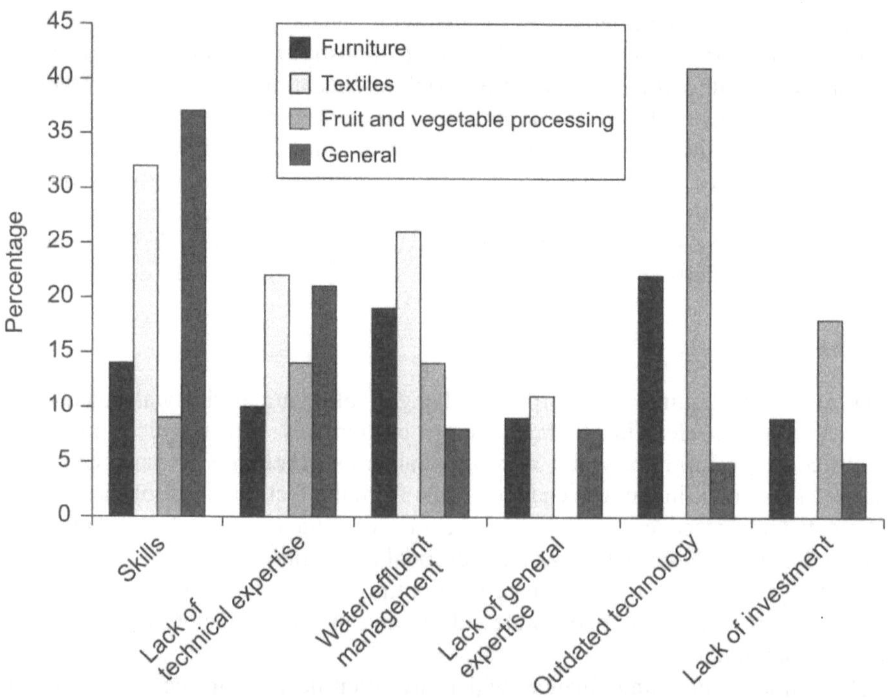

Fig. 7.3. Technological problems faced by SMEs in the UK/ROI

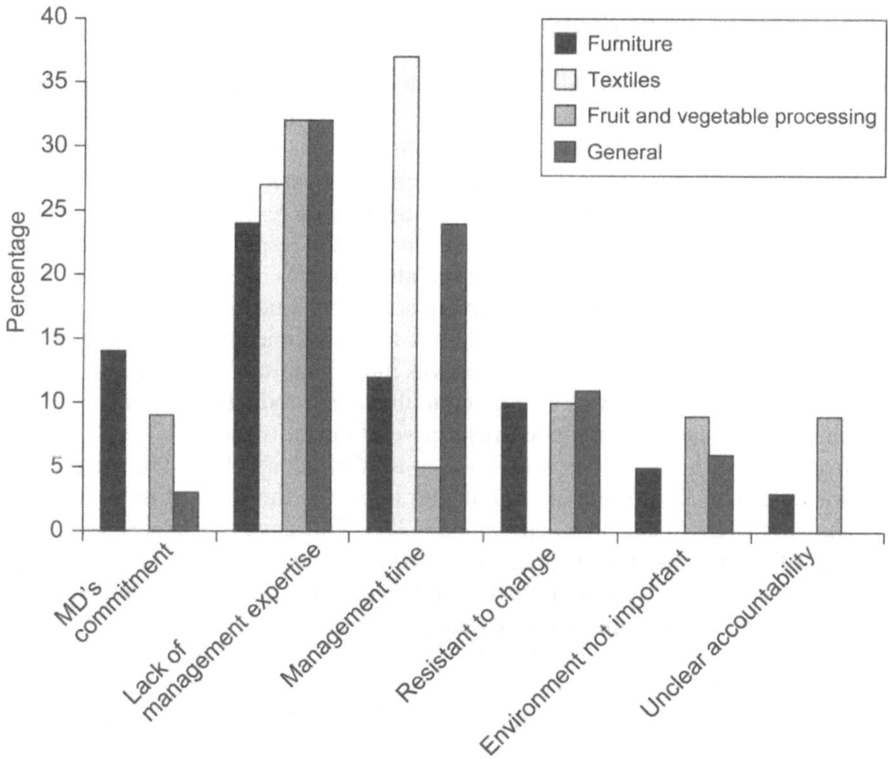

Fig. 7.4. Organisational problems faced by SMEs in the UK/ROI

Typical technological problems varied from a lack of proper skills (at the shop floor level) and expertise (management level), lack of technical expertise, outdated technology, problems encountered in effluent management, lack of sufficient investment to undertake suggested changes. The typical organisational problems were a lack of commitment, time and expertise on the part of management; absence of a responsible person to cater to the environmental needs of the firm, attitudinal problems including a resistance to change and an unwillingness to give environmental issues a priority.

While in Germany and Italy there were no clear technological or organisational problems, interviews with the advisors in the UK/ROI revealed a range of technological and organisational problems experienced by SMEs (Figs. 7.3 and 7.4).

7.6 Importance of Advisory Services

To measure the efficacy of the advisory process, the SOAs were asked to comment on whether their services led to a significant reduction in the environmental impact of client firms in the selected industries.

7.6.1 Impact of the advisory services

Most interviewees claimed that their advice substantially contributed towards a change in the firm's environmental position. This may have been the result of implementing an Environmental Management System (EMS) or other input-orientated systems, which created information and awareness of adopting various initiatives. Some of the benefits realised by SMEs as a result of introducing EMS were cited as: greater awareness of environmental responsibilities among the workforce; better relationships with regulatory bodies, insurers and the local community and better internal management and communication. It was also stated by some advisors that for a substantial shift in environmental attitude the person responsible for environmental initiatives was more important than the initiative itself. The opposite opinion was also given in that the introduction of EMS will lead to a shift in environmental attitude irrespective of the individual.

Most UK/ROI interviewees stated that in all the industrial sectors, there was a considerable reduction in the environmental impact in the form of reduced waste (45%), energy (29%) and VOC emissions (13%). This gave rise to cost savings as a result of firms adopting their advice. However, there was no prominent improvement in some of the environmental indicators like management and water efficiency as a result of adopting the advisory services.

Of the total German interviewees, 83% observed an environmental improvement after an advisory input. The main developments were in the field of energy and material input, which were significantly reduced after upgrading machinery or implementing a controlling system. On the output side, developments were through end-of-pipe technologies for cleaner waste-water, reduced soil impact and processes involving reduced noise pollution (this was especially true in the textile sector). These improvements accrued along the whole lifecycle of the products. As a result of the consulting process some firms experienced mixed results regarding cost effects. Examples were quoted of firms who had reported an increased labour requirement to implement the suggestions and some quoted increased productivity after setting the proposals into place.

In terms of investments, which follow the advice, much end-of-pipe technology reduced the environmental impact. Examples were cited of investments in IT systems as a common approach to managing the information input necessary to run an EMS following an advisory input. These measures also led to increased running costs e.g. employing a project manager or updating electronic systems, but they were not substantially higher than before. Few of the advisors interviewed trained firm employees on special technologies that may have to be adopted due to special product characteristics e.g., specific filtering or packaging technology.

In Italy, the majority of the sample (89%) observed that advisory services could be effective. In particular, they were considered to be effective in bringing about a long-term lowering of the physical impact on the ecosystem (waste reduction, quality of emissions). As a result of seeking advice many firms in Italy, realised an increase in production levels (24%) significant cost reduction (21%) and improvement in work quality (14%). On the other hand the impact of the advisory services on the firm's culture was less visible due to the limited goals of the advisory sources and the fact that financial concerns within the SMEs received a higher priority.

Increased economic efficiency, as an important measure of the advisory activity of SMEs was also investigated. Though most of the SOA involved in production and technical changes in SMEs were unable to give figures, in the UK/ROI and in Germany almost all those interviewed and about half of the Italian advisory firms acknowledged that their involvement led to a significant increase in firms' economic efficiency.

7.6.2 Importance of firm culture in the adoption of clean technologies

Three strategies were identified where management may influence the attitudes of SMEs towards being more pro-environmental.

- Managers can influence the organisational attitude towards the environment while improving co-ordination inside firms and act as mediators between different constraints (production, financial and environmental concerns).
- They can influence SME's attitudes through decision-making, encouraging changes and innovations that have positive effects on the environment.
- They can promote cultural change while modifying the organisational attitude towards ecological issues, for example by including protection of the environment among the organisational goals expressed or implementing a reward system for improvement in ecological standards.

Information was also sought on whether management or firm culture had an impact on the successful adoption of environmental initiatives. Management attitude was largely believed to play an emphatic role on the environmental performance of firms across all the countries. This is contrary to the findings (except for the Italian fruit and vegetable processing sector) of the attitudinal analysis undertaken and discussed in Chap. 8.

Conclusions therefore could be drawn that cultural aspects arise with respect to the difference between successful and unsuccessful implementation of advice. Innovative instruments that go beyond compliance with regulations were implemented in firms where there was a progressive management team. One of the German advisors thought that a family-based firm offered the chance for co-operation because of the owner's close business involvement but it may also be resistant to change because of his/her traditional behaviour. A further factor was the financial status of the SME. If firms thought that the financial position was sound, it was easier to achieve implementation of advice. In general, an environmental friendly culture is necessary to go beyond regulation and the most influential person to promote such initiatives is the owner or director with his/her leading position in the firm.

7.6.3 Best sources of advice (as perceived by the SMEs)

The manufacturing firms sampled were asked what they thought were the best sources of information on environmental legislation and green technologies. Responses from the firms in the UK/ROI indicated that professional associations were the best sources for obtaining information on environmental legislation and green technology. A significant percentage of the firms also felt that public bodies were good sources of information on environmental legislation. In the case of Germany, professional associations were considered to be the best sources of information for environmental legislation while most firms felt that fellow companies were also useful sources of information on green technology. In Italy, the majority of the firms felt professional associations were the best sources of advice for both environmental legislation and green technology. Surprisingly environmental associations ranked low as a preferred source among the countries (Table 7.2)

Table 7.2. Best sources of information identified by the firms, percentage of firms

	UK/ROI		Germany		Italy	
	EL	GT	EL	GT	EL	GT
Fellow companies	16.9	11.8	18.4	28.4	9.0	11.3
Professional associations	39.6	37.3	48.4	17.3	82.8	72.1
Private consultants	17.2	13.3	5.4	14.0	7.0	9.2
Public bodies	22.4	19.6	19.3	7.4	–	6.2
Environmental associations	3.9	11.6	5.2	6.2	–	–
Others	–	6.4	3.3	26.7	1.2	1.2

EL Environmental legislation, *GT* Green technology.

Most SMEs seek cost efficient and speedy solutions to their problems. The ability of most professional associations to respond to the firms' problems with a solution quickly, through some of their unique services (for example in the UK/ROI, associations operated 24–48 hr response services and helplines) gives them an edge over other advisory services. Although some firms in the UK/ROI stated that professional advisory services were very specialised and at times this had worked against them in that the professional associations were sometimes unwilling to deal with issues what were, in their opinion, trivial matters.

In the face-to-face interviews, the manufacturing firms were asked to list the best advisors for information on clean technology. In the UK/ROI, manufacturers considered government bodies and regulators to be the most important source of advice. Other important sources of advice were consultants, suppliers of machinery and industry associations. A sectoral analysis for UK/ROI revealed that industrial associations and suppliers feature as important advisors in the furniture sector. In the textile sector, consultants followed by regulators/government bodies and suppliers were voted as important advisory sources, while in the food sector, regulators/government bodies followed by customers and consultants were important (Fig. 7.5). However, during the interviews almost all the sources of advice, maintained that they were the best source of advice as they catered to all the advisory demands of the SMEs.

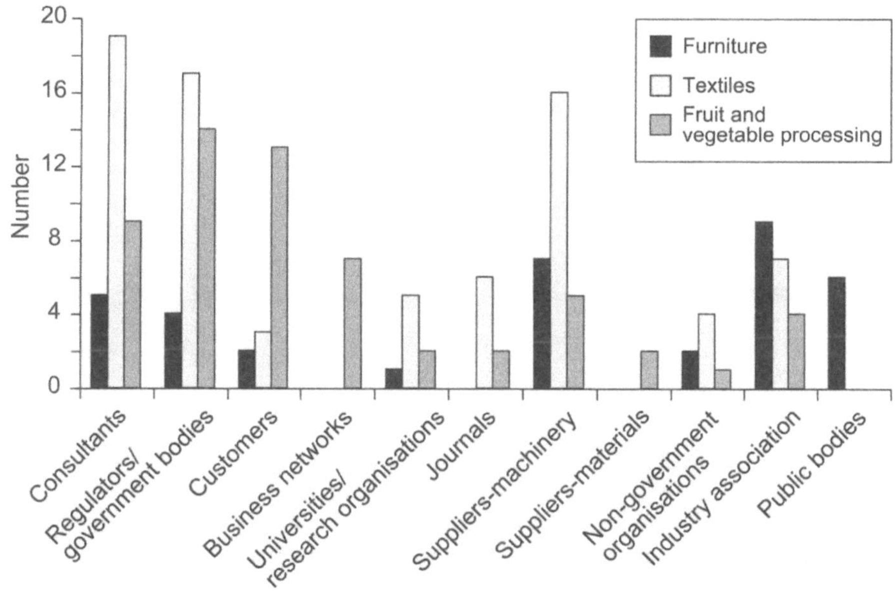

Fig. 7.5. Important advisory sources identified by SMEs in the UK/ROI

In Germany, the situation was somewhat different with 37% of the SMEs regarding suppliers of machinery and materials as the best sources of information. A similar percentage of firms indicated that journals/magazines and customers served as advisory sources of quality. As regards each sector, suppliers, Internet and journals/magazines were considered to be the best sources of information in the furniture sector. In the textile sector it was customers, suppliers and journals/magazines while in the fruit and vegetables sector it was journals/magazines, suppliers and industrial associations (Fig. 7.6).

In the case of Italy, there was insufficient data for the furniture and textile sectors but in the food sector the main sources of advice as identified by the SMEs were the Internet followed by consultants. Suppliers of machinery and materials, unlike in Germany and the UK/ROI, did not seem to play a major role as advisory bodies.

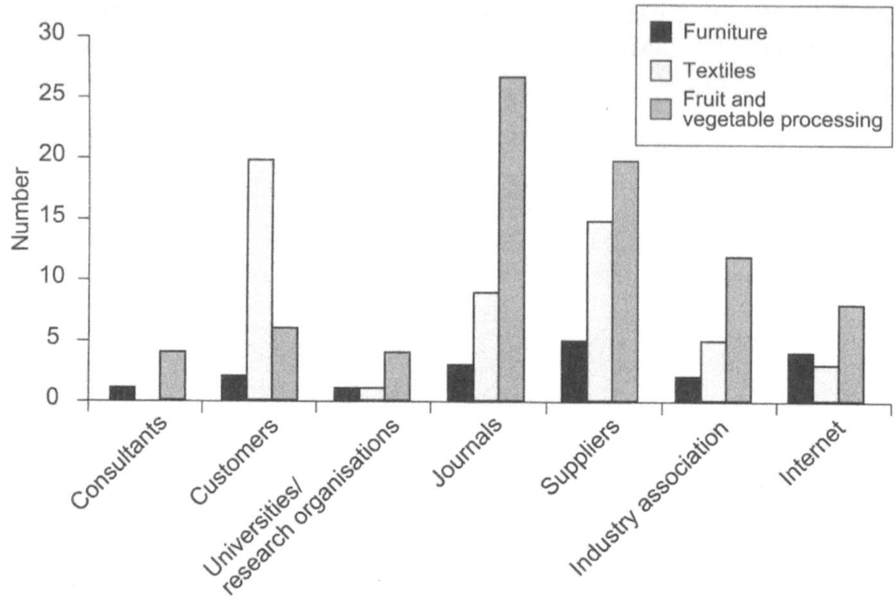

Fig. 7.6. Important advisory sources identified by SMEs in Germany

7.7 Competition between Private /Public Agencies

From the interviews it was clear that there was competition to provide advisory services between public agencies and non-profit organisations, and private consultancy companies. In the UK/ROI, there was some degree of competition regarding advisory services for waste, water and energy management. To maintain market position advisory organisations were continually upgrading their services and seeking to differentiate the services they provide from those of their competitors. This involved improving/modifying the in-house design and project management capability with access to substantial technical and operational resources, which would appeal to a broad client base.

In Italy, competition with private or public agencies existed in certain areas especially information services for sector specific issues, implementing environmental guidelines or aiding in the preparation of technical reports. The Italian experiences stated that the reason for the competition was because of the strict requirements in environmental standards, thus, creating increasing business opportunities for private companies operating in the 'green' sector. Also, the scarcity of public provision and services (waste disposal, handling of hazardous materials, hygiene and safety of products) in some regions favoured the spread of a number of micro-enterprises specialised in niche services (recycling and assembling of wastes).

In Germany, consultants faced some competition with the public agencies; for example, some municipalities have started advising local production firms about waste management. These are new competitors, which have developed in recent years because of a new service orientation in public administration.

Both in the UK/ROI and Germany, it was reported that universities were considered as the main competitor as regards information sources for quality management and measurement techniques. However, they also observed that private advisory firms achieved better quality in their processes, because universities can often be too theoretical in their approach. Public administration and professional associations are very specialised in their advice and may not take into consideration apparently insignificant elements, which may in fact be a contributory factor towards the problem.

In Italy, respondents evaluated the consultant's professional knowledge as an advantage over the competitors, while other factors of differentiation included the cost of intervention and capacity/willingness to provide long-term assistance to clients. Technical specialisation in different sectors was also considered to be relevant.

7.8. Role of Universities and Research Institutions

Sources of advice from universities and research organisations in the UK/ROI and Germany were interviewed to ascertain the role they played in influencing and disseminating information on the take up of clean technology by the SMEs. The range of services offered to the small business community included:

- A basic site-based environmental assessment to benchmark the company's performance and to identify issues that will improve their environmental performance and identify issues that can be resolved in-house or through projects.
- Access to expertise within the organisation to meet specific environmental or business requirements of SMEs.
- Opportunities for company-specific research projects.
- Literature featuring updates on current legislation and new technology, ideas and suggestions for good practice.
- Offering access to a comprehensive information resource including electronic and paper-based information on a range of environmental topics.
- Raising environmental awareness in suppliers through supplier development events organised by the centre.
- Conducting workshops and seminars on environmental issues, which provide opportunities to network with potential customers, suppliers and organisations offering help and guidance on a range of environmental issues.

- Other services include a range of training and academic courses at reduced cost; opportunities for research and development into new technologies and products and environmental audits.

Though some of the services provided are generic, innovative new services like gap analysis, advice on integrated management systems and accreditation advice are unique to most research organisations. These institutions were of the opinion, that the new services ensured the attraction of new clients and the retention of existing ones. In spite of the range of services offered, some organisations indicated that it was very difficult to involve SMEs in the programme. They were of the opinion that, smaller SMEs were unwilling to change because they anticipated a cost factor and the structure of most of these firms did not support the adoption or implementation of environmental initiatives. Hence, most research centres maintained a superficial involvement with the SMEs.

The major gap in poor interaction was the inability on the part of the research centre to respond promptly to the firms' problems with a solution. The prevalence of many competitors and the duplication of services could be attributed to the ineffective flow of information. However, the reputation of research organisations in providing quality service played a major role in attracting firms to seek advice and in this way they fared better than their competitors.

Problems encountered by research organisations in the implementation of environmentally relevant innovations and providing advisory services to the SMEs were similar to those encountered by the consultants irrespective of country. Prominent among these were cost, existing corporate culture (also in the sense of old-established processes and techniques) and lack of proper knowledge and awareness about environmental issues among SMEs. It was pointed out that educational insufficiencies represented one of the major difficulties in attracting potential clients; likewise price considerations play an important role that could not be underestimated.

The research institutions interviewed were unable to estimate the implications of their proposals on the client company's sales, costs or labour productivity. However, many interviewees attributed the effects of their service, to an increase in productivity, reduced costs and increased customer satisfaction. Significant improvement in the environmental indicators, for example, considerable reduction in waste, reduction in energy use, awareness of the impact of climate, issues of hazardous waste and recycling waste were also mentioned as being benefits of the advisory service.

Views expressed by some of the organisations revealed that the basic managerial attitude towards ecological questions, internal communication and the existing corporate culture played a vital role in the adoption of advice. On a technical level, managerial staff and the staff who have received training exerted the biggest influence on the successful implementation of initiatives.

7.9 Role of Suppliers

The actual need to change can be influenced by the supplier or customer. Suppliers of materials and equipment were interviewed. Manufacturing firms are often regular clients of the supplier and visited regularly by technicians or salespeople. The supplier may offer training courses about new products/machinery. The operating costs associated with new machinery comprise cost for energy, water, steam, etc. and it can be difficult for the supplier to quantify the potential gains. It was emphasised that difficulties frequently arose during the starting period of newly delivered machines due to lack of proper skills in the factory, which is why most suppliers offered specified training to the client's operating staff.

Equipment suppliers contact potential customers by various means. Many approached their potential clients by active advertisement or presentation of the product range at various trade fairs or selling their products through sales representatives. Some suppliers in the UK/ROI and Germany stated that sometimes, potential customers contacted them directly often because they needed assistance as a result of new regulations. Although, the ultimate aim of the process is the sale of the product in question rather than the environmental advice, most of those interviewees said that it was an effective way of creating or raising environmental awareness among SMEs. Almost all suppliers interviewed emphasised the fact that their products have a positive environmental effect. They asserted that they were using the best available technology in terms of ecology. The environmental improvements achieved were stated to be improvement/purification of air and the reduction of the amount of energy, water, and chemicals used. The interviewees believed that their work influenced their client's approach to ecological questions in a positive way.

Typical problems in attracting clients to seek help or advice from a supplier involved the projected investment, the potential client's willingness to change and to bear certain risks. Firms that are most successful in adopting changes are those with a *quality* culture and commitment by top management.

In Germany, customers were identified as being important providers of advice for SMEs in the textile finishing industry. In this sector there was direct business-to-business environmental communication between the textile finishers and the clothing manufacturers they supplied. There was a lack of this type of inter-business relationship regarding environmental issues in the furniture and fruit and vegetable sectors.

Textile finishers supply clothing manufacturers who in turn supply retailers with the final product. Hence, environmental communication is an integral part of the supply chain. Additional research in Germany examined the extent and importance of the instruments used in environmental communication between the textile finishers and the clothing manufacturers.[1] In general, it was found that product-orientated communication, such as product data sheets and Eco-tex labelling was the main form of communication employed with process-orientated communica-

[1] Detailed information is available from Michael Keil at IÖW.

tion i.e. communication about the environmental impact of the production process itself, only being important for clothing manufactures that placed a strong emphasis on supplying environment friendly clothes.

It was to the advantage of both the finishers and clothing manufacturers that long-term co-operation was established. Clothing manufacturers stated that they were prepared to help their suppliers (textile finishers) even if they failed to adapt to changes, which were considered important to the clothing company. The rationale for this was based on the belief in the supplier's competence on a general basis and also the difficulties in finding new suppliers that could deliver quality products. The clothing manufacturers are usually large companies with substantial managerial resources, which can be utilised by the textile finishers to provide managerial advice.

7.10 Role of Professional Associations

As explained in the earlier sections of this chapter, interviews were carried out with representatives from the public and professional advisory bodies across all the four countries. While these bodies were dispersed uniformly in the UK/ROI and Germany, in Italy they were clustered in the northern regions, where most of the Italian SMEs are concentrated, and where compliance with environmental norms is much higher. In the regions of the South and Centre, where the economic stability of small firms also depends on the opportunity to minimise costs related to environmental norms, compliance is less frequent and public advisory services offer basic services and concentrate on environmental information campaigns.

The type of advice given to SMEs in Italy depended more on the geographical location of the firms. Firms in Northern Italy tend to seek issue-specific advice e.g. information on the toxicity of products, energy reduction, use of new technologies and emission reduction. While in the southern regions most centred on demand for waste management and normative compliance. In particular, public advisors in the food sector tend to work mostly with problems related to safety and hygiene of products and in some cases with the use of eco-labelling. Non-profit organisations working in the furniture sector, paid much attention to recycling issues, such as the reuse of wooden materials and the reduction of toxic coatings.

In Italy, waste treatment and collection tended to be the primary target of all the organisations, especially for the agencies located in the South and Centre. In the Northern regions, more emphasis was placed on technology use and innovation. Advice provided by agencies of the local Chambers of Commerce is oriented to improve all the different facets of the SME's activities (from sales to use of energy) and to encourage SMEs in implementing new environmental production practices (ISO 14000 or EMAS).

Territorial differentiation was not observed in either the UK/ROI or Germany. Professional associations offered consultancy services, applied R&D, training and issue-specific solutions irrespective of their location. Involvement with the firm tended to be quite intensive, usually involving 2–3 year programmes. However, some advice was dealt with over the telephone, for example, legislation queries.

Lack of internal expertise, use of advisors in the past and unwillingness to change were cited as the main problems in persuading clients to take advice. These problems were more prevalent in smaller companies. Again in the view of the professional associations management culture is very important and if changes are to be successfully implemented top management need to be more open to change and have a long-term vision.

7.11 Key Findings

Most of the advisors interviewed argued that their services definitely helped in reducing the environmental impact. The role of management was critical in the environmental performance of the firms and that management generally sought to obtain short-term solutions rather than consider the long-term implications of existing problems.

In Germany, most SMEs sought free advice and felt that the quality of service offered by public bodies and private consultants were not very different. In Italy, SMEs observed that publicly provided services were more reliable and efficient than those of private consultants. In the UK/ROI, SMEs were prepared to pay for advice but considered that there was little to differentiate the quality of service between public and private providers.

7.11.1 Issues specific to the UK/ROI

The major problem identified in the Irish sample was that SMEs were resistant to change. They were very resistant to external advice on change unless the firm had initiated the decision to seek change. The resistance to change stemmed from the fact that most firms are family run, with low craft and managerial skills. They produce very traditional furniture and sell to a declining market. The advisors felt that many of these firms were not run along commercial lines, with family considerations overriding good business practice.

Some key observations made by the SOA on the textile sector, were:

- the disaggregated supply chain works against environmental improvement,
- the SMEs' resistance to change, and their lack of knowledge and experience to define a problem,
- the textile sector is old, more rigid and focuses on traditional operational methods.

In the fruit and vegetables processing sector in Northern Ireland the lack of professional management in family owned businesses, an excess of advisory sources and the dilemma faced by the SMEs in choosing the most appropriate advice for their specific needs were highlighted.

7.11.2 Issues specific to Germany

Some issues were identified by advisors, which were not specific to any one sector but in general applied to all sectors. Lack of knowledge about the importance of the problem was a major issue. There were often problems with the acceptance of information by SMEs. Limits to the activities of advisors are set by the ability of the firms to take up advice. Imperfect information did exist in cases where new unproven technology and information about future policies was unclear.

In small firms, or not-so-professional family companies, a typical organisational problem was observed. Often there were no written responsibilities, which could be litigated, thus no one has the responsibility to ensure the implementation of the advice. In addition, lack of time and/or human resources, which are necessary for a successful commitment to the process, were problematic. There was a general perception that some family firms only "do what they always have done" and were not willing to change their behaviour. Unwillingness to accept advice was seen as a typical problem in Germany among the smaller firms. Thus it is not surprising that, on average approximately 15% of the clients were unsuccessful in the application of the results from the advisory process.

7.11.3 Issues specific to Italy

The issues identified were not sector specific but can be referred to all sectors. Smaller firms were often unwilling to change, since they were more worried about short-term survival and were unlikely to see the benefits from environmental initiatives. Management is too busy dealing with day-to-day activities and found it difficult to consider long-term strategies. Further, the inability to have a vision was compounded by economic constraints and lack of human resources (in many cases a quality manager or a production manager played the role of an environmental officer as well).

Public inspections and control on firms' activities, however, have still a relevant role in preventing major problems such as territorial degradation, and in reducing air and water pollution, as well as in raising work safety standards. Nevertheless, public and private advisory services in the environmental arena, frequently complemented one another and provided SMEs with qualitatively different benefits and evaluations.

7.12 Conclusion

There was much variation across country in terms of the process of advice delivery and the need for SMEs to have advice on environmental issues. In terms of problem identification, the majority of German and UK/ROI SMEs approached the advisory services with specific problems while in Italy the advisory services as well as the SMEs together identified any existent problems. The duration taken to provide an effective solution varied across country depending on the gravity of the problem and the solutions often suggested implementing changes in the production process.

In terms of the importance of take up and use made of the external sources of advice, advisors in all the countries agreed that information sources were available in plenty but not effectively taken up by SMEs. Although in Italy, the advisors did recognise that small firms were unable to search for information in the market, probably due to a limited knowledge about the issue.

The abundance and easy access to information could be reinforced by the fact that none of the manufacturing firms pointed to a lack of sufficient advice as an obstacle to the adoption of clean technology, except in the case of UK's furniture sector. The compliance only firms in this sector felt the lack of advice as an impediment to adopt clean technology.

Possible reasons for the reluctance on the part of SMEs in all the industrial sectors to seek advice from the advisors were cited as follows:

- There were instances where the SMEs had received conflicting advice from the advisory services furthermore each supplier offers their own, unique system, making it difficult for firms to assess what is right.
- Past bad experiences of inadequate/inefficient advice received from advisory sources by SMEs. One of the advisors interviewed was quoted as saying "use of advisors in the past and the purchase of unsuitable technology based on the advice received, are both factors that can lead to difficulties in persuading clients to take advice". The advisors, in general, did not feel this was a problem except for one advisory source that confirmed that firms find it very difficult to get access to advice on environmental technology and management.

Most SMEs remained unaware of the environmental impacts associated with their business activities and had a low awareness of environmental programmes or resources available to them. This general lack of concern for the environment was compounded by the problem of unclear and ambiguous environmental terms and concepts. Organisational constraints, lack of time and resources were major problems; the latter reduced the firm's capacity to solve problems in-house and also reduced their ability to anticipate problems before they became constraints. A major obstacle common to all industrial sectors and countries was the unwillingness to change. In Germany, lack of capital to incorporate changes following an advisory input seemed to be a significant problem, while in Italy the cost of advisory services was recognised as a major constraint.

Very small firms were the least able to implement change. The main problems with these firms was that they were often family owned, handed down from the previous generation, managed by the owner who may lack the skills and entrepreneurial qualities needed.

Small and medium sized enterprises can be very resistant to the idea of using consultants, and were more comfortable with the services of university-based advisors. They were also sceptical about the extent to which opportunities for cost savings existed and rarely could be persuaded by cost-savings alone as a reason for undertaking environmental projects. The role of environmental champions to drive through change was seen as very important. As there is clearly no single best way to implement change, it would be best to have a plurality of ways of promoting/supporting change and so advice providers should understand different ways of responding. While the medium sized firms were more concerned about the payback period, the smaller firms need more counselling than they ask for; since smaller firms were usually not aware of what initiatives can be carried out and how this could be done.

Other major obstacles to adoption of clean technology were lack of resources, firm culture, company strategy and use of technology. The human resource constraint was, in general, more important than financial constraints. The smaller the firm the more tasks have to be undertaken by one particular person. For example the health and safety manager, the production manager or quality manager may have the job of addressing environmental issues. Time to consider and time to implement is very important and although SMEs may not explicitly calculate the costs of their time they were very aware of time constraints.

The resource constraint is combined with firm culture within the firm, namely the view that environmental concerns are a necessary evil, and dealt with through compliance. The result is that environmental responsibility is often bundled with health and safety and not viewed as a separate issue. This reinforces the low priority, and often leads to environmental improvements being treated as 'interruptible processes', that take second place to production related activities. Firms do not look for the most efficient way of managing environmental concerns, but instead they focus on compliance. Environmental performance is not viewed by firms as being directly related to competitiveness, therefore the corporate culture has an important role in supporting and sustaining the firm's commitment to continuous environmental improvement.

Increasingly, smaller firms are beginning to focus on cost-minimisation as a competitive strategy rather than solely on sales maximisation. Margins are being squeezed by increases in input costs that cannot be passed to final prices because of tight product markets.

The UK/ROI has a problem with technology that is outmoded and therefore has a high environmental impact. Industry is not interested in optimising technology and appears often to be unable to modernise. While in Germany there is a culture of respect for technology as the foundation of the firm, in the UK/ROI the culture is more that of technology as a servant of the firm. SMEs are isolated, they may not have the networks to understand and evaluate technology. This places the power with the suppliers who are able to tell firms what they need. Another feature that emerged was that the UK also suffers from a lack of a proper system of skill development for new/young workers. This reinforced the view among many SMEs that "business isn't about learning, it's about doing business".

German experiences stated that within a given firm there might be resistance to change. Firms may not be open to change and management may have an obstructive agenda. Often health and safety managers were insufficiently powerful to be able to push though the environmental improvements that are required. The culture of the firm may have more impact than the size of the firms when it comes to change and champions are vitally important in a firm that does not have a culture of innovation. In Germany, firms were more aware of their responsibilities towards environment but the lack of capital and resistance to change especially on the part of the smaller firms was seen as a major hindrance.

Quality of advice was stated as problematic in both the public and private sectors. There were problems with poor credentials and also with quality of staff and service delivery. Some British advisory services felt that there was a tendency for firms to view government as a problem for business. Public sector advice is grant driven especially for SMEs, but it was apparent that the reason for the decreased range of services/programmes available to firms was due to the fact that some government advisory sources were of the view that free service can be abused and the return may be poor.

A stronger partnership between the public and private sector providers of advice needs to be established. The public sector needs to accept that they alone cannot supply all demands and that better results can be achieved by using a partnership approach. SMEs often consider the adoption of an EMS as a knee-jerk reaction to a specific problem, such as supply-chain pressures. They are not able to take a strategic or long view, as they do not have the necessary planning skills, visionary outlook or they cannot understand the link between economics, the environment and the social impact. Since many SMEs do not regard the environment as an important issue, they do not feel the need for any advisory services. Radical changes may be needed which could include a combination of committed top management working with the support and involvement of the entire firm. A senior manager may act as a project champion, giving weight to the project. Firms that are successfully able to change are firms where the project champion has the vision to see beyond the day-to-day constraints. Another approach may be the adoption of a two-pronged strategy of increased pressure on SMEs to adopt *eco-efficiency* as a fundamental business principle, coupled with programs that simultaneously increase environmental awareness.

As suggested by some of the advisory services, the local Chamber of Commerce could play a major role in educating and providing useful information to the SMEs. There were suggestions from the advisors e.g. imposing a mandatory but subsidised membership fee to all the industries in the region, which would benefit the Chamber of Commerce as well as the industries in enabling them to procure useful information and networking of services.

For successful implementation it is important that there is top management commitment and awareness. This will affect the priority that the firm places on environmental impact. Another important factor is that firms do not view systems as being merely bureaucracy but are aware of the added value they provide. If a firm has a positive experience of implementing quality systems this will increase their openness to environmental systems.

8 Cultural Attitudes

8.1 Introduction

In the last decade, the incidence of ecological problems caused by the activity of firms has grown dramatically. This has raised a number of questions among economists and social scientists regarding the need to consider environmental issues within the production process and an understanding of the organisation's view on the operation of firms. However, despite the fact that the focus has been mostly concentrated on the level of economic relationships, an increasing number of scholars regard the "environmental problem" primarily as one of social structure and social economic organisation (Egri and Pitfield 1996).

Integrating the economic approach with an analysis of cultural and institutional aspects of socio-economic activities enables consideration of the problem from a wider perspective. Further it allows us to understand the activity of firms in terms of the joint product of economic and socio-organisational aspects (Egri and Pitfield 1996). From a sociological perspective, firms do not perform in narrow environments as autonomous entities; on the contrary, they interact with other influences in society, authorities, consumers and political organisations. In fact, organisations and firms are *embedded* in different kinds of environment (Granovetter 1985), either in the form of internal human and technological resources or in the form of ecological and socio-cultural surroundings. This interaction contributes to a shared influence that has a relevant impact on the firm's action and attitudes.

A number of possible factors have been suggested as indicators or determinants of the environmental performance by SMEs. One of which have been the attitudes of the owner/manager, since all individuals approach environmental problems with their own set of pre-existing values and assumptions. In most SMEs, management and ownership of the firm are frequently synonymous. As a result, the owner/manager has a major impact on the performance of the business (Storey 1994). This also provides a unique opportunity for any individual to put their values into practice in the workplace, and to influence the behaviour of employees, consumers and other stakeholders.

Gerrans and Hutchinson (1998), Merritt (1998) and Ludevid (2000) have shown that many SME owners/managers have a high awareness of their role in environmental remediation, and have a desire to take action. However, most studies have reported that there is often a great deal of difference between the stated attitude of owners/managers and the actual environmental performance of the firms. Tilley (1998), Hutchinson and Chaston (1994) and Petts et al. (1999) have all reported a similar dichotomy between attitudes and environmental performance amongst SME owners/managers in the United Kingdom and continental Europe. This gap has been termed the *SME problem* in environmental management (Merritt 1998; Tilley 1999).

The approach that firms use to deal with environmental issues is a growing area of contemporary management research (Fineman 1997; Welford 1995). The management of environmental issues is important for any firm irrespective of firm size. Hodge (1995) has pointed out that the resolution of environmental problems requires analysis and understanding of '…..incentive and choices at the very microeconomic level.. Why do individuals and firms make particular choices and how can we persuade them to act differently? Even at this scale, the factors are many and complex' (p.10). A similar view was taken by Cramer (1998), who claimed that '…little research has so far been conducted into how companies put environmental management into practice and the internal and external forces that influence them' (p.163). Petts et al. (1998) argued that the relationship between behaviour and attitudes, and the discrepancy between what managers say and do about environmental matters, needs more investigation and explanation.

The present research hypothesised that the internal 'culture' of the firm is important in effecting environmental decision-making and environmental performance. The significance of the role of top management is important and an understanding of how managers become convinced to choose cleaner technologies given the competitive position of the enterprise.

8.2 Measuring Cultural Attitudes

The cultural attitude analysis was carried out based on the responses received from respondents to an attitudinal questionnaire. Table 8.1 lists the number of responses received from each country and industrial sector.

Table 8.1. Responses to the cultural questionnaire

	UK/ROI	Germany	Italy	Total
Furniture	25	29	33	87
Textiles	25	29	33	87
Fruit and vegetable processing	26	34	30	90
Total	76	92	96	264

The cultural questionnaire was designed to help determine management attitudes. Using Likert scales, the relationship of the firm's attitude was measured, for example, a) the degree of pro and anti environmental attitudes b) the relationship between the environmental strategy adopted and the attitude towards the environment c) environmental concerns d) government policy towards the environment and its effect on the individual firm e) organisational culture and structure etc. In addition, views about the role of science and faith in modern life, the effects of human interference on natural environment and the relationship between economic growth and environmental protection were considered as a part of the analysis. Attitudes were measured with a number of interrelated questions partly as a check on the consistency of attitudes held and also to link attitudes to environmental initiatives undertaken by the firm.

As discussed in Sect. 2.3.1, four questions (Q1 to Q4) in the cultural questionnaire played a key role in determining the attitude of the firm. A cultural score was calculated based on the degrees of agreement or disagreement indicated by each firm's response to 15 statements from this set of 4 questions. For each statement a maximum score of 4 and a minimum of 1 was allotted. The sum of scores for the 15 statements made up the cultural score. Firms with a cultural score of more than 45 were classed as *pro-environmental* and firms with a score less than 45 were classed as *anti-environmental.*

8.3 Pro and Anti-environmental Attitudes

Table 8.2 shows the extent of firms adopting a pro-environment attitude across the samples. For the furniture sector, a higher proportion of firms in the UK/ROI were pro-environment than their German and Italian counterparts. The percentage of compliance plus firms that were pro-environment was slightly greater in the UK/ROI than in Germany, where an equal proportion of compliance plus and excellence firms were pro-environmental. In Italy, a majority of the compliance only firms were pro-environmental compared to compliance plus firms.

Table 8.2. Percentage of firms adopting a pro-environmental attitude

	UK/ROI	Germany	Italy
Furniture	68.0	65.5	49.9
Textiles	60.0	45.0	69.6
Fruit and vegetable processing	65.0	56.0	38.0

In the textile sector, compared to the UK/ROI and German firms, a greater percentage of Italian firms were pro-environment, of which, compliance plus firms were more pro-environmental in comparison to the other performance groups. In the UK/ROI, a greater percentage of compliance only firms were pro-environmental while in Germany, a greater proportion of pro-environmental firms belonged to the compliance plus and excellence group.

As in the furniture sector, there were a higher proportion of pro-environmental firms in the fruit and vegetable processing sector in the UK/ROI, compared to Germany and Italy. On average, more compliance plus firms were pro-environmental in the UK/ROI, while in Germany, there was little difference between compliance plus and excellence firms. In Italy a majority of the excellence firms were pro-environment compared to the other performance groups. However no statistically significant relationship could be established between the attitudes and the environmental performance of the firms.

On the whole, across the industrial sectors and the countries, compliance plus firms were more pro-environmental except in the case of the UK/ROI's textile firms. However, a statistical significance was not established between the groups.

8.4 Environmental Strategy and Environmental Attitude

The cultural scores were compared with the environmental strategy of the firms in all three sectors across sample countries. Tables 8.3 to 8.5 present the average cultural scores for pro and anti-environmental firms adopting various environmental strategies.

For the furniture sector, more pro-environmental firms in the UK/ROI adopted an environmental strategy focused on regulation, whilst the pro-environmental firms with a high cultural score adopted either a strategy of focus on environmental performance or on eco-products. In Germany, pro-environmental firms with a high cultural score adopted an environmental strategy of regulation and eco-products. None of the Italian firms had a score greater than 45 so could not be classed as pro-environmental.

Table 8.3. Environmental strategy and environmental attitude, furniture sector

Environmental strategy	Average cultural score Pro-environment firms			Average cultural score Anti-environment firms		
	UK/ROI	Germany	Italy	UK/ROI	Germany	Italy
Regulation	47	46	–	39	39	33
Eco-efficiency	47	–	–	39	38	41
Env-performance	49	–	–	–	36	–
Eco-products	50	46	–	39	43	–
New eco-markets	–	–	–	–	–	–
Public perception	–	–	–	–	–	–
Total	48	47	–	39	38	34

In the UK/ROI textile and finishing sector, comparatively more firms that were pro environmental adopted an environmental strategy driven by regulation. However, firms having the highest cultural score adopted a strategy of using environmental performance to differentiate them. In Germany, firms with the highest cultural score adopted a strategy focusing on eco-products, while in Italy they adopted a strategy that focused on new eco-markets (Table 8.4).

Table 8.4. Environmental strategy and environmental attitude, textile sector

Environmental strategy	Average cultural score Pro-environment firms			Average cultural score Anti-environment firms		
	UK/ROI	Germany	Italy	UK/ROI	Germany	Italy
Regulation	50	–	45	37	37	36
Eco-efficiency	45	–	–	35	37	37
Env-performance	63	48	–	38	36	39
Eco-products	–	52	–	44	31	32
New eco-markets	–	–	45	–	–	36
Total	50	50	45	37	36	36

For the fruit and vegetable processing sector in the UK/ROI, a greater number of pro-environmental firms adopted an environmental strategy with a focus on eco-efficiency. In Germany more pro-environmental firms adopted an environmental strategy of eco-efficiency and eco-products, while in Italy more pro-environmental firms adopted an environmental strategy focussing on eco-efficiency (Table 8.5).

Table 8.5. Environmental strategy and environmental attitude, fruit and vegetable processing sector

Environmental strategy	Average cultural score Pro-environment firms			Average cultural score Anti-environment firms		
	UK/ROI	Germany	Italy	UK/ROI	Germany	Italy
Regulation	47	45	46	41	35	38
Eco-efficiency	47	46	46	43	30	44
Env-performance	45	–	46	38	42	44
Eco-products	55	47	46	–	36	–
New eco-markets	–	45	–	–	44	–
Public perception	49	–	–	–	–	–
Total	48	46	46	40	35	41

In summary, the relationship between environmental strategy and environmental attitude varied across the countries and the sectors, and there was no clear evidence of a significant relationship between the environmental strategy adopted by firms and their environmental attitudes.

8.5 Concern for the Environment

From Sect. 8.3, it was clear that there was a concern for the environment as evidenced by the pro-environmental stance taken by many compliance plus firms. The level of importance for this concern can be judged to a certain extent by examining the views expressed by the respondents about the various strategies by which government could encourage firms to use less energy and produce less waste. Firms were asked to respond to the following question in terms of degrees of agreement or disagreement with the statements.

What in your opinion was the main impact of environmental legislation?

a. An incentive to improvement for the good of society.
b. A hindrance to market competitiveness.
c. Encourages increased efficiency of production.

Responses to this question allowed a three-way classification of firms as *pro-environmental*, *anti-environmental* or *innovator*. In this case the cultural score did not determine the class, i.e. whether a firm is pro or anti environmental, instead the statements asked gauged the attitude. Pro-environmental firms were those that considered legislation as an incentive, leading to improvements that benefited society as a whole while anti-environmental firms were those that consider legislation as an impediment to overall competitiveness. The third category, innovators, considered such legislation as leading to higher efficiency through increased cost savings or market opportunities.

In the UK/ROI, more of the compliance plus firms in the furniture sector attributed the main impact of environmental legislation to the improvement of the good of society in general, while more of the compliance only firms in Germany and Italy did so. A reasonable proportion of firms in all the countries, associated the impact of environmental legislation with encouraging increased efficiency of production, hence they were innovator firms. In general, few firms indicated that environmental legislation operated as an obstacle to market competitiveness (Table 8.6).

Table 8.6. Percentage of firms categorised on the basis of opinion regarding environmental legislation

	Innovators	Pro-environment	Anti-environment
Furniture			
UK/ROI	12	68	16
Germany	24	41	28
Italy	21	57	18
Textiles			
UK/ROI	24	56	20
Germany	21	34	45
Italy	33	54	12
Fruit and vegetable processing			
UK/ROI	27	50	23
Germany	41	21	32
Italy	80	20	–

In the textile sector, firms across the sample countries indicated that the main impact of environmental legislation was to improve the good of the society in general. It was however notable that a substantial percentage of German firms were of the opinion that environmental legislation operated as an obstacle to market competitiveness.

As regards the fruit and vegetable processing sector, in the UK/ROI, more of compliance only firms attributed the main impact of environmental legislation to the improvement of the good of society, while it was more of the compliance plus or excellence firms in Germany and Italy that did so. Again as observed in the textile sector, a reasonable proportion of firms across the countries associated the impact of environmental legislation with encouraging increased efficiency of production.

The average number of initiatives adopted (as measured by eco-points) by pro-environmental, anti-environmental and innovator firms, is shown in Table 8.7.

Table 8.7. Average number of initiatives (eco-points) adopted by attitude to environmental legislation

	Innovators	Pro-environment	Anti-environment	All firms
Furniture				
UK/ROI	3.88	6.58	5.91	6.28
Germany	18.38	14.94	16.00	15.89
Italy	6.75	6.07	5.16	5.86
Textiles				
UK/ROI	10.54	5.44	8.53	8.53
Germany	15.83	18.03	17.53	17.37
Italy	6.18	3.56	3.75	4.45
Fruit and vegetable processing				
UK/ROI	12.19	10.25	10.50	11.15
Germany	17.18	17.76	14.81	15.95
Italy	13.82	7.71	–	12.60

There was no significant difference in the environmental performance between the pro-environmental and anti-environmental firms in the three industrial sectors. German innovator firms in the furniture sector took up, on average, more initiatives than the other two groups. In the UK/ROI and Italian fruit and vegetable processing sector, innovator firms undertook, on average, more initiatives than the other two groups of firms. The reason is because innovator firms are often product or process innovators and this is reflected in their adoption of new technologies.

The trend in the results shows no clear indication of any positive relationship between good performers and pro-environmental attitudes among the surveyed firms. Assuming further, that the innovator firms are also pro-environmental, then by reclassifying the firms into only pro and anti-environmental categories Table 8.8 shows that there was little difference in the average number of initiatives undertaken by category.

Table 8.8. Average number of initiatives (eco-points) adopted

	Pro-environment	Anti-environment
Furniture		
UK/ROI	6.18	5.91
Germany	16.21	16.00
Italy	6.17	5.16
Textiles		
UK/ROI	9.01	8.53
Germany	17.20	17.53
Italy	4.55	3.75
Fruit and vegetable processing		
UK/ROI	10.93	10.50
Germany	17.37	14.81
Italy	12.59	–

The level of concern for the environment among the performance groups varied across the sectors and the countries sampled. However, there was a certain degree of consistency in the textile sector across the countries, where firms indicated that the main impact of environmental legislation was to improve the good of society in general. There was no clear indication of a positive effect of culture on the environmental performance of the firms.

8.6 Degree of Environmentalism

Having determined the level of the firms' concern for the environment, the degree of environmentalism was judged from the responses received asking why the firm had adopted environmental initiatives. Respondents choose from:

1. My company has adopted environmental initiatives because of legislation.
2. My company has adopted environmental initiatives spontaneously because it is in the interest of all to live in a clean environment.
3. My company has adopted environmental initiatives because of new types of consumption now fashionable among our customers.
4. Others (specify).

For the furniture sector, the answers to this set of statements show that a majority of German firms had adopted environmental initiatives because it is in the interest of everyone to live in a cleaner environment, demonstrating they were strong environmentalists. A larger percentage of firms in the UK/ROI and Italy indicated that legislation was the main driver for adopting environmental initiatives, thus exhibiting a much lower degree of environmentalism (Table 8.9).

Table 8.9. Degree of environmentalism (as % of firms)

	Strong	Medium	Compliance	Other
Furniture				
UK/ROI	20	8	48	24
Germany	72	14	13	–
Italy	9	–	82	3
Textiles				
UK/ROI	32	4	52	12
Germany	35	21	42	3
Italy	51	3	42	3
Fruit and vegetable processing				
UK/ROI	27	19	46	8
Germany	73	9	15	3
Italy	47	37	17	–

In the case of the German textile firms, there was an almost equal distribution between the various degrees of environmentalism, indicating the same emphasis on clean environment, market and regulation. In the UK/ROI, 52% of the textile firms show a lower degree of environmentalism by stating that they adopt environmental initiatives because of regulation. In Italy, 51% of the textile firms showed a strong environmental position while 42% showed a low degree of environmentalism.

Responses from the fruit and vegetable processing firms show that a majority of German and Italian firms had adopted environmental initiatives because it is in the interest of everyone to live in a cleaner environment, showing, they were strong environmentalists. A greater percentage of firms in the UK/ROI indicated that legislation was the main driver for adopting environmental initiatives. Also a significant number of UK/ROI firms and Italian firms pointed out that market was a major driver to adopting environmental initiatives.

8.6.1 Benefits of adopting environmental initiatives voluntarily

While it is clear from the above section that most German firms exhibited strong environmentalism, UK/ROI firms lower environmentalism and Italian firms exhibited both strong and low environmentalism. Firms were asked whether they profited by adopting environmental initiatives voluntarily (Table 8.10). Except in the case of Italian compliance only firms, most furniture firms in the UK/ROI and Germany felt they did not accrue any benefits by adopting environmental initiatives voluntarily. In the textile sector, the UK/ROI and German firms stated benefits to be 'low' or 'moderate'. In the fruit and vegetable processing sector, more firms in the UK/ROI and Germany stated that they accrued, little or no benefit at all. Italian firms on the other hand stated they benefited from adopting environmental initiatives voluntarily.

The results for Germany are not consistent with the results obtained in the previous section, where a majority of German firms, exhibited a strong environmental position. However, a firm conclusion cannot be drawn because the questionnaire did not ask for the nature or extent of benefits accrued by the firm.

Table 8.10. Benefits of adopting environmental initiative voluntarily (as % of firms)

	High	Moderate	Low	None
Furniture				
UK/ROI	–	12	8	76
Germany	–	3	17	80
Italy	94	3	–	–
Textiles				
UK/ROI	4	24	32	40
Germany	–	28	20	52
Fruit and vegetable processing				
UK/ROI	–	2	42	46
Germany	8	6	12	71
Italy	33	47	–	3

No data was available for the Italian textile sector.

8.7 Government Policy

Firms were asked to state their agreement/disagreement to the following set of statements on what were the various strategies that government could encourage companies to save energy and produce less waste.

1. Investing public money in information campaigns to encourage companies to save energy/waste reduction.
2. Implementing a program of increasing taxes on energy consumption/setting limits on the use of packaging for products.
3. Increasing taxes for companies which use excessive amount of energy/ not collect products at the end of their lifecycle.
4. Reducing taxes for companies with low consumption production process/companies using recycled materials.

Analysis of the responses indicated that generally firms in the UK/ROI preferred that public money is invested in information campaigns to encourage firms to save energy and produce less waste. In Germany 31% and in Italy 18% were in favour of such a strategy. This strategy was more favoured in the fruit and vegetable processing sector in the UK/ROI and Italy compared to the other sectors (Fig. 8.1).

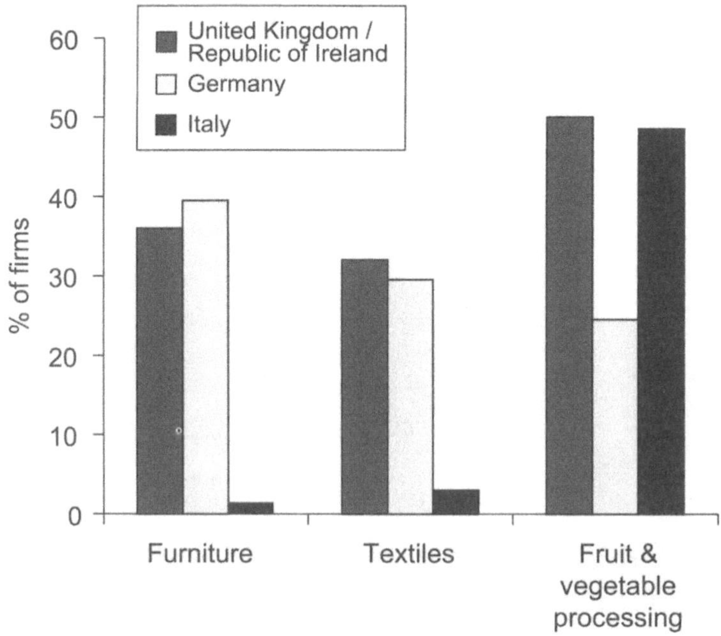

Fig. 8.1. Percentage of firms agreeing strongly with investing public money to encourage companies in energy saving and waste reduction

German firms in general strongly favoured taxing companies that generate more waste and utilise excessive energy compared to the UK/ROI and Italian firms. An analysis by sector for each country is shown in Fig.8.2.

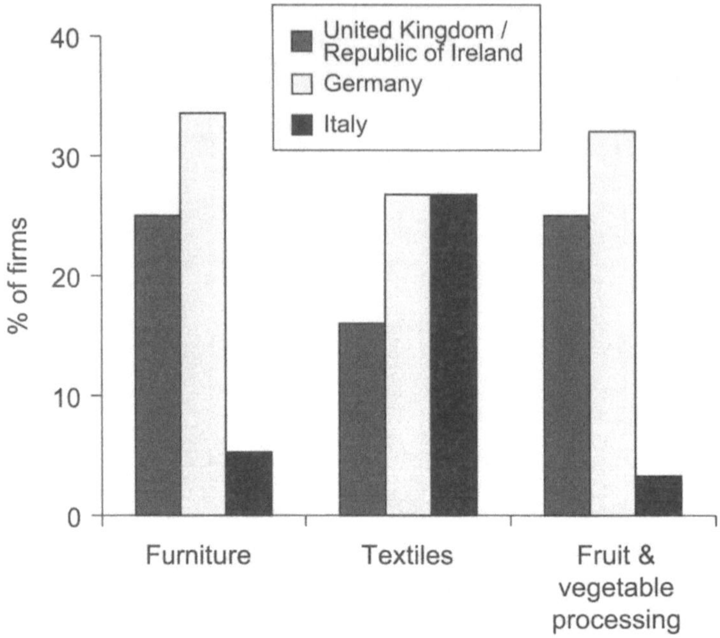

Fig. 8.2. Percentage of firms agreeing strongly with increasing taxes for firms consuming excessive energy or producing more waste

All three sectors in Germany strongly preferred to encourage firms to use less energy and produce less waste by taxing those who use more energy or produce more waste. This kind of response in Germany may be due to the wide acceptance of the eco-tax system. In the UK/ROI, views expressed on this issue by the furniture and fruit and vegetable processing sector were stronger compared to those of the textile firms. While in Italy, textile firms seemed to be more in favour of taxing firms compared with firms in the other two sectors.

When views about the reduction of taxes for firms using less energy and producing less waste was gathered, it was observed that 76% of the Italian firms were in favour of this strategy, followed by 51% of UK/ROI firms and 43% of German firms. Analysis of the responses by sector shows that, on average, a greater percentage of furniture and fruit and vegetable processing firms in all the countries in favour of a reduction in taxes for firms using less energy and producing less waste compared to textile firms (Fig. 8.3).

The results of the analysis show that firms in the UK/ROI and Italy prefer a governmental policy that is less harsh compared with Germany where firms are most strongly in favour of raising taxes on polluting firms.

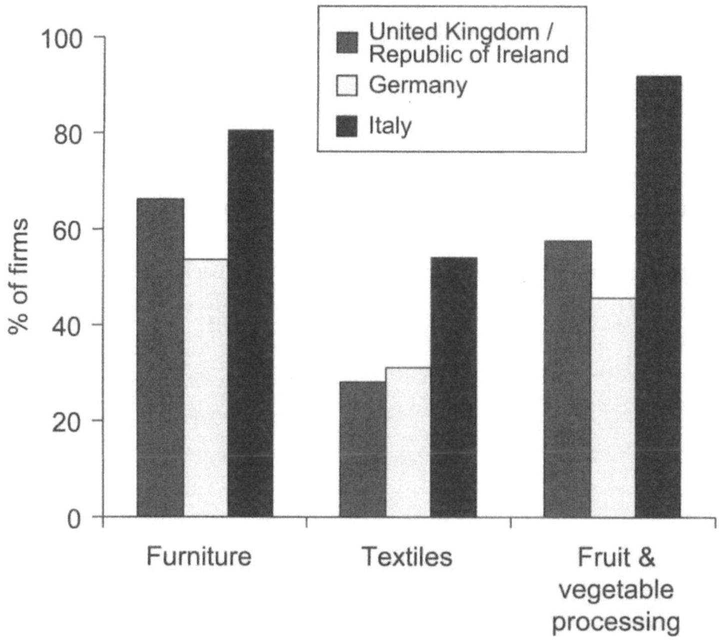

Fig. 8.3. Percentage of firms agreeing strongly with reducing taxes for firms consuming less energy or producing less waste

8.8 Firm Attitudes, Environmental Performance and Size

8.8.1 Environmental performance

Analysis indicated that pro-environmental firms were better environmental performers (based on the evidence that more of the compliance plus firms had a pro-environmental attitude). However, with the exception of the Italian fruit and vegetable processing sector, there was no correlation between environmental attitude and environmental performance. Hence, the cultural attitudes of the firms did not have a direct influence on the adoption of environmental initiatives. Firms adopted environmental initiatives for reasons such as regulation or the market rather than an actual concern for the environment. Though in the case of Germany more firms seemed to adopt environmental initiatives due to a general concern for the environment. However, a statistically significant relationship between attitude and performance could not be established with the better environmental performing firms

8.8.2 Firm size and environmental attitudes

The relationship between size and environmental attitudes was examined for firms in each sector across the sample countries. Table 8.11 presents the average cultural score for firms in each size group for the pro-environmental firms. There was no evidence of a size effect on the environmental attitude of the firms.

Table 8.11. Average cultural score of pro-environmental firms in each size group

	0-10	11-25	26-50	51-100	101-250	251-500	Total
Furniture							
UK/ROI	52	45	–	46	47	46	48
Germany	46	46	–	–	50	46	47
Italy	–	–	–	–	–	–	–
Textiles							
UK/ROI	53	–	45	–	52	48	50
Germany	–	52	–	–	48	–	50
Italy	45	–	45	–	–	–	45
Fruit and vegetable processing							
UK/ROI	46	51	47	–	47	–	48
Germany	46	45	–	–	–	47	46
Italy	–	46	47	46	46	–	46

8.9 Environmental Orientation

In order to gauge the environmental orientation of the firm respondents were asked the extent to which they agreed or disagreed with each of four statements. Statements 1 and 2 can be thought of as representing an *anti-science* environmental attitude, involving scepticism about the role of modern science and a belief that human intervention and economic growth inevitably harm the environment. Statements 3 and 4 can be considered to represent *modernist* beliefs, a sense that too much fuss is made about the environment and that economic imperatives should always come first (Table 8.12). The two sets of statements are not mutually exclusive. An interviewee may exhibit both an anti-science attitude as well as a modernist view of the environment. Table 8.13 shows the percentage of firms that strongly agreed with each statement.

Table 8.12. Environmental orientation

Anti science statements	Modernist statements
1. Economic development cannot be supported by the natural resources available	3. People worry too much about the fact that economic development damages the environment
2. It is right to increase taxation for industries which pollute	4. Nowadays, full employment is more important than environmental protection.

Table 8.13. Environmental orientation, percentage of firms strongly agreeing with each statement

	UK/ROI			Germany			Italy		
	Fur	Tex	F&V	Fur	Tex	F&V	Fur	Tex	F&V
Anti-science statements									
1.	12	16	31	62	52	50	9	18	–
2.	56	56	31	45	17	17	3	36	–
Modernist statements									
3.	–	16	4	7	34	15	6	3	17
4.	8	20	4	21	31	9	–	–	37

Fur Furniture, *Tex* Textiles, *F&V* Fruit and vegetable processing.

In order to examine the relationship between the two extremes, an attitude scale (*pro-science* attitude scale incorporating statements 1 and 2 and an *anti-modernist* scale combining statements 3 and 4) was devised. The scale was constructed by calculating the average score based on the respondent's answers to each statement. The maximum score for each scale was 4 (strong disagreement with *both* statements in the scale) and the minimum score was 1 (strong agreement with *both* statements in the scale). Thus a low score in the pro-science scale was equivalent to an anti-science environmental orientation, while a low score in the anti-modernist scale conveyed a modernist orientation.

Table 8.14. Pro-science and anti-modernist scores

	UK/ROI	Germany	Italy
Furniture			
Pro-science	2.85	3.20	2.10
Anti-modernist	2.90	2.80	2.70
Textiles			
Pro-science	3.05	2.60	2.70
Anti-modernist	2.30	2.05	2.80
Fruit and vegetable processing			
Pro-science	2.90	2.90	2.55
Anti-modernist	2.75	2.90	1.80

The scores on the pro-science scale and the anti-modernist scale (Table 8.14) were, on the whole, above the average (2.5). This signified that, on average, firms held neither an anti-science view (involving scepticism about the role of modern science) of the environment nor a modernist view (i.e. that there is too much fuss made about the environment).

8.10 Organisational Structure and Culture

The cultural questionnaire included questions that enabled us to determine the organisational model the firm followed and hence investigate whether there was an association between the organisational structure and the environmental attitude of the firm. The firms were also grouped into three culture categories groups based on their responses to a set of statements.

8.10.1 Organisational Structure

By examining the level of responsibilities within the firm an insight into the organisational structure existing in the firm can be obtained. Using the following question, firms were categorised into belonging to one of a possible three models.

Which of the following principles do you believe to be more important for the success of your firm?
1. That employees conscientiously fulfil the tasks assigned to them.
2. That the best technical solution is always the target.
3. That individual responsibility is taken at each level of production.

Organisations having an *authoritarian* model (point 1) would generally have a low regard for environmental issues while firms with a *participatory* model (point 3) would be at the opposite end of the spectrum. A neutral or mid-way environmental position was held by firms following a *technocratic* model (point 2). Table 8.15 shows the percentage distribution of the sample firms adopting each of these models.

In both the furniture and textile sectors, over half of the firms in the UK/ROI and Italy adopted a participatory structure, (generally analogous with good firm culture). In Germany, furniture firms were almost equally divided between the participatory model and the technocratic structure (associated generally with medium environmental attitude). Textile firms, in Germany, showed a fairly high proportion (32%) adhering to the authoritarian structure. In the fruit and vegetable processing sector, firms in the UK/ROI and Germany adopted a either a participatory structure model or an authoritarian model while in Italy firms were equally divided between all three models. On the whole, firms were inclined to adopt a participatory structure, which is associated with a good firm culture. However, interviews with the various sources of advice (see Chap. 7) revealed that a top management commitment to the environment is important in enabling firms to adopt cleaner technology and this is normally based on an authoritarian style of leadership.

Table 8.15. Percentage of firms in each organisational model

	Participatory	Technocratic	Authoritarian
Furniture			
UK/ROI	54	42	4
Germany	45	48	7
Italy	58	12	30
Textiles			
UK/ROI	74	17	9
Germany	42	26	32
Italy	69	28	3
Fruit and vegetable processing			
UK/ROI	65	–	35
Germany	59	9	32
Italy	33	37	30

Likert (1961), stressed the fact that freedom for groups to set their own work pace produces a high degree of professional commitment, so long as there is sufficient interaction between the individual, his colleagues, and his superior. He states that 'the individual must be part of an active social system where there is frequent contact and interaction. This interaction motivates the individual'. Therefore participative management requires a *link pin* function whereby the group's attitudes are represented and communicated in both directions within the hierarchy. This *link pin* assumes a leadership role and helps to create an atmosphere of co-operation among group members, prompting them to interact with one another. Likert describes the leader's role as follows: "The leader fully reflects and effectively represents the views, goals, values and decisions of his group in those other groups where he is performing the function of linking his group to the rest of the organisation. He brings to the group of which he is the leader the views, goals and decisions of those other groups".

However, it is worth noting that the qualities essential to leadership in participative management are quite different from those required in an autocratic one; the leader, instead of taking decisions for his or her subordinates, expresses and communicates one group's view to the other.

8.10.2 Organisational Culture

Firms were asked to respond to the following question that enabled us to determine their culture.

To what degree do you personally agree with each of these statements?

1. Company culture is formed by the values of the director/owner, management, shared by his/her assistants and employees.
2. Company culture depends upon the marketplace in which the company works.
3. Company culture is a combination of customer, employees and management values.

Based on the responses, firms were allocated to one of three categories. *Mixed culture,* where culture was a combination of customer, employee and management values. *Shared culture,* where company culture was formed by the values of director/owner and management. *Embedded culture*, where company culture depended on the market place in which the company operated. Table 8.16 shows the percentage of firms in each of these three culture categories.

Table 8.16. Percentage of firms in each culture category

	Mixed	Embedded	Shared
Furniture			
UK/ROI	28.0	12.0	60.0
Germany	51.7	13.8	34.5
Italy	45.5	18.2	36.4
Textiles			
UK/ROI	36.0	16.0	48.0
Germany	31.0	20.7	48.3
Italy	34.4	25.0	40.6
Fruit and vegetable processing			
UK/ROI	42.3	26.9	30.8
Germany	44.1	14.7	41.2
Italy	70.0	10.0	20.0

The majority of the furniture firms in the UK/ROI indicated a shared culture while in both Germany and Italy a mixed culture dominated. In the textile sector, in all the countries, there was a preference for a shared culture closely followed by a mixed culture where all stakeholders could make an impact. In the fruit and vegetable processing sector, the vast majority of the Italian firms indicated a mixed culture while in Germany and the UK/ROI the preference was for a mixed approach closely followed by a shared culture. It was apparent that the embedded culture was not very widespread in any of the industrial sectors or countries sampled.

Analysis revealed that manufacturing firms in the sampled sectors across the countries, on average, adopted the participatory organisational structure. This highlights the fact that the firms believed that the success of the firm was based on the individual responsibility taken at each level, reflecting a good firm culture. A similar trend was observed regarding organisational culture in the fruit and vegetable processing sector, where the majority of firms adopted a mixed culture. This probably was due to the nature of the fruit and vegetable processing sector which is primarily market driven.

8.11 Conclusion

One of the objectives of this research was to test the hypothesis that culture is important in an organisation and the role played by top management is important for an understanding of how firms are convinced to choose clean technologies given the competitive position of the firm. With this view a detailed attitudinal questionnaire was set out and firms were asked to respond. The cultural questionnaire was designed to help determine management attitudes, which were measured for example by determining:

- The degree of pro-environmental attitudes and the concern for environment
- The relationship between the environmental strategy adopted and attitude towards environment
- Government policy towards the environment and its effect on the individual's firm
- The influence of organisational culture and structure.

In addition views about the role of science and faith in modern life, the effects of human interference on natural environment and the relationship between economic growth and environmental protection was considered as a part of this analysis.

Results of the analysis show that attitudes towards environment were on an average positive. A greater number of respondents in the sample countries voiced their concern about environment. Also their concerns emanate in the responses to some of the questions asked to them. However, there is little relationship between these attitudes and environmentally responsible behaviour of the firms who were interviewed. Personal attitudes of the firms did not translate directly into environmental performance. In most of the SMEs, whilst recognition of the importance of the environment was high, the capacity to act was low because there has been no sustained pressure to improve.

The majority of the sources of advice interviewed in all the countries agreed that management attitudes could play a vital role in the adoption of clean technologies. However, analysis of the responses reveal that environmental attitudes did not have any significant impact on the environmental performance of the firms, except in the case of Italian fruit and vegetable processing sector, where a significant correlation between environmental performance and environmental attitudes could be found.

There was a prevalence of shared culture in the textile firms across the sample countries and in the UK/ROI furniture firms. Sources of advice (see Chap. 7), also strongly believed that culture and the role of management are very important, in the adoption of environmental initiatives. Fairbairn (1997) pointed out that a role culture, which is epitomised, as the traditional hierarchical structure wherein the firm culture is formed by the values of the director/owner of the firm is perhaps the most appropriate culture. He further stated that its strengths are in its predictability and stability while its obvious opposites of inflexibility and slowness of reaction are its weakness. For the implementation of an EMS such as ISO9000 this structure has many advantages, and it can be argued that ISO9000 was originally designed for this type of environment with its layers of procedures and instructions mapping the general hierarchy.

As discussed earlier, Likert (1961), suggests that a *link pin* in a participative management can assume a leadership role and help in creating an atmosphere of co-operation within a firm, further, prompting them to interact with one another. This suggestion invariably emerged from the interviews we had with the sources of advice.

Differences in the preferences of the respondents, for example, firms were more likely to favour reducing taxes for environmentally friendly firms than increasing taxes on environmentally harmful behaviour, were observed in the UK/ROI and Italy. While in Germany, respondents were more in favour of increasing taxes on polluting firms. These national differences may be accounted for by variations in educational and income levels between the countries, or differences in the extent to which some environmental issues have spread beyond the concerns of the elite. Previous work has shown that such background factors are important in shaping environmental attitudes (Witherspoon and Martin 1992; Ester 1981; Heath et al. 1991). However, Schaper (2002) points out that demographic variables e.g. age gender and educational level are not related to firm's actual performance. On the other hand key external variables e.g. amount of time available for owners to undertake discretionary business activity and the level of environmental information available to business owners shows a significant positive relation to a firm's 'green' activities. In this research it did emerge that lack of management time was a constraint on the part of SMEs in the take up of cleaner technologies, however enough research could not be carried out to link this constraint with the firm's pro-environmental attitude

From the analysis of the environmental orientation of the firms it is evident that firms are not sceptical about the role of modern science nor feel that there is too much emphasis given to the environment. A complex picture emerges out of this research, on management culture and environmental performance. There are positive attitudes towards the environment. These differ between countries. There is only scattered evidence of a relationship between positive environmental attitudes and environmental performance. There is a difference between the stated attitude of owners/managers and the actual environmental performance of the firms. More detailed analysis at the microeconomic level is therefore required to explain these patterns of behaviour.

9 Summary and Conclusions

The major aim of this research was to identify and weigh the importance of factors that promote and constrain the adoption of environmental initiatives by SMEs.

9.1 Focus of Study and Hypotheses

The study focused on the characteristics and experience of those above average environmental performers in three industrial sectors namely, furniture, textile finishing and fruit and vegetable processing, and compared their characteristics with average performers in the same industry.

The study sought to test three key hypotheses:

1. The relationship between investment in environmental initiatives and firm competitiveness was likely to involve positive feedback in both directions.
2. The role of management and the culture of the business organisation were important to the take up of environmental initiatives.
3. A third focus of the research pertained to the influence of information and advisory sources in that the dissemination of information about environmental initiatives was likely to be a critical determinant of diffusion.

The study recognised a group of sub-hypotheses especially related to the first point above, in particular:

- Modern machinery can embody good environmental performance.
- Skills will foster initiatives.
- R&D will enable firms to take up technologies.
- National regulation and grants foster environmental performance.

It also recognised that there were a wide set of other barriers which research measurement does not focus on, including return on investment, preference for end of pipe technology, risk, capital rationing etc. While these additional barriers were brought within the research framework, they were not measured.

The relevance of the research was primarily to help increase the focus of both industrial and environmental policy instruments, to improve their effectiveness and efficiency in obtaining goals like "sustainability" and a better integration of industrial/regional policy at the national and EU levels with environmental policy.

9.2 Summary of Findings

This section summarises the findings from the research and draws together conclusions from the industries sampled and country comparisons.

9.2.1 Environmental Performance

Furniture

On the basis of a simple count Italy undertook most initiatives followed by Germany then the UK. For the UK/ROI and for Germany there was evidence of a relationship between numbers of initiatives adopted and firm size. There was no relationship in the case of Italy in which most sample firms were, in any case, very small.

Textile finishing

German sample firms undertook more environmental initiatives followed by the UK then Italy. Generally there was a relationship between size of firm and environmental performance.

Fruit and vegetable processing

German firms undertook substantially more initiatives, followed by the UK/ROI and Italy. There was a statistically significant association between size of firm and numbers of initiatives adopted for the UK/ROI and Germany but not for Italy.

Conclusions on overall environmental performance

With the exception of furniture where Italian firms showed on average the best environmental performance, German firms were ahead in the other two industries, the UK/ROI and Italy followed with less differentiation between them. In Germany and the UK/ROI there was a relationship between size of firm and the number of initiatives adopted but this was less evident for Italy. In the case of the latter this may reflect the smaller average size of firm sampled and the less differentiated driver. Regulation predominated in Italy, for example, while the market was especially important for German firms.

9.2.2 Hypothesis 1

The relationship between investment in environmental initiatives and firm competitiveness was likely to involve positive feedback in both directions.

Analysis of this hypothesis involved a set of tests to relate firm competitiveness with firm environmental performance. There were four tests. The first considered a set of variables that reflected overall firm performance (including output and input measures of firm performance). The second analysed the economic impact of individual initiatives, and combinations of those initiatives. The third considered obstacles to the adoption of individual initiatives. And the fourth examined the relationship between the environmental strategy adopted by the firm and the economic or competitiveness strategy adopted, and the firm's environmental performance.

I. Environmental performance and measures of firm performance

The key competitiveness variables included: size (economies of scale); productivity; employment growth; destination of sales; R&D; age of machinery; labour force qualifications.

Furniture

There was limited evidence to support the hypothesis that the more competitive the firm the more environmental initiatives it would undertake, although there were inconsistencies between the two surveys (face-to face interviews (FTF) and postal questionnaire (PQ)) and between countries. For British and Irish firms there were statistically positive relationships with productivity, age of machinery, and employment growth. For Germany there was only a relationship with export propensity. In the case of Italy there was a small correlation with exports.

Textile finishing

A relationship between environmental performance and firm competitiveness based on measures of R&D, firm growth, and exports showed no relationship for the PQ data except between exports and environmental performance in Germany. Similarly the FTF data showed no significant differences with the exception of the age of machinery, and R&D in Germany. Hence there was little support for the hypothesis.

Fruit and vegetable processing

In Germany and Italy but not in the UK/ROI, there was a relationship between numbers engaged in R&D and environmental performance. There was no relationship between environmental performance and firm growth for any of the countries. For Germany and Italy but not the UK/ROI there was a relationship between export propensity and environmental performance. Hence, based on the PQ data, there was some evidence to support the hypothesis that environmental performance is related to competitiveness. Additional measures used in the FTF survey showed no relationship between environmental performance and productivity, age of machinery or employee qualifications, though there was some indication on performance and age of machinery for Germany.

Conclusions on findings on competitiveness

There was clearly no convincing evidence that firms with a better economic performance adopt more environmental initiatives. The evidence was scattered and inconsistent between countries to suggest any policy conclusion that can be derived from the findings e.g. that the presence of R&D supports environmental improvements. The evidence was also too weak to draw distinctions between industries and countries. What can clearly be noted, however, is that the firm's competitive position is not a hindrance to environmental performance or that the take up of environmental initiatives weakens the competitive performance of firms.

II. Individual environmental initiatives and firm performance

In this analysis consideration was given to (a) the pattern of adoption of environmental initiatives as environmental performance rises, and hence the overall economic impact; and (b) drivers (e.g. cost, regulation, market etc.) and the economic impact (on cost, profits, employment etc.) of individual environmental initiatives.

Furniture

Moving up the environmental hierarchy (from compliance only to compliance plus to excellence) involved a progression from housekeeping initiatives to additional more technically demanding initiatives to the inclusion of management initiatives. Movement along this hierarchy, in the UK/ROI alone, involved increasing firm size. For the other two countries, size was not related to environmental performance. Compliance only firms focused on waste separation and undertook virtually no management related activities. Compliance plus firms were involved in waste separation, chemical control and energy efficiency. Excellence firms were distinguished by the almost universal participation (except for Italian sample firms) in management initiatives.

Motivation to undertake the initiatives included regulation and cost to reduce waste (embracing the reuse of packaging), cost savings to improve energy efficiency, regulation for solvent reduction. The market was an important driver for the avoidance of toxic substances in raw materials, eco-design and environmental management systems (although there were variations, for example, UK/ROI firms reported that regulation and health and safety were the two drivers influencing the adoption of their environmental management systems). The use of environmental management systems were a function of the size of the firm.

What was the economic impact of these initiatives? From the firm's position the overall impact of increased sales and reduced costs with associated increased productivity signalled a rise in economic performance. For waste reduction, reuse of packaging, lightweighting, use of wood burning boilers, the experience was positive, generally cost reducing, though the experience was mixed. Each of these is driven by cost but also regulation. Eco design, environmental management, use of solid wood, which are driven primarily by the market had cost raising implications but generally improved sales.

The experience of impacts was mixed. In the case of purely regulation driven initiatives, solvent reduction and water based coating, the experience was mixed according to whether the outcome of the adoption of the initiative increased or reduced costs, compared with the former method. The overall economic impact was mixed: positive in the FTF survey but more pessimistic in the PQ survey.

The ratio of positive to negative effects across the different impacts on sales, labour, costs and profits showed much similarity between compliance only and compliance plus firms, with on balance compliance plus firms experiencing slightly better economic performance as a result. By size, firms in the smallest group, under 25 employees reported more negative impacts than larger firms. The overall impact on use of labour was positive.

Textile finishing

As in furniture, as firms moved up the scale of environmental performance they graduated from solely housekeeping type initiatives to activities that required technical expertise, to organisational and market focused initiatives. Moreover small firms favoured waste management initiatives and as size increased proportionately more firms undertook efficiency, market and management initiatives.

Comparing compliance only firms with compliance plus and excellence firms there were significant differences in the types of environmental initiatives undertaken. Compliance plus firms undertook more initiatives in energy, water protection, chemical substitution, packaging, environmental management and communication than compliance only firms in the UK/ROI. While packaging initiatives were ubiquitous. Similarly in Germany compliance plus performers undertook significantly more management, chemicals, energy and water initiatives, while excellence groups undertook more management systems. There were no comparable data for Italy.

Controlling chemical inputs was in general costly and was regulation driven in the UK/ROI and market and regulation driven in Germany. Air protection measures were motivated by regulation and firms used end of pipe technologies. For water use and protection, improvements were made for regulatory and cost reasons. Environmental communication and management were importantly driven by the market in Germany and regulation in the UK/ROI. As for furniture, waste initiatives were motivated by regulation and cost, and energy initiatives by cost.

There was some variability in economic impacts for the same initiative within and between countries. In the UK/ROI the aggregate of economic impacts indicated that cost driven initiatives accounted for 77% of economic impacts. In Germany they accounted for a third, the remainder were predominantly accounted for by market driven initiatives.

The net effect across all initiatives was clearly positive with respect to the impact on profits, and while there was variability, the effects were much more clear-cut than for furniture.

Comparing the ratio of positive economic impacts to initiatives between compliance, compliance plus and excellence performers, the data showed that economic benefits rise through the performance groups. This was not clear in the case of furniture. The relationship with size indicated fewer benefits for the very small firms. There was a small gain in labour accounted for by the initiatives.

Fruit and vegetable processing

In the furniture and textile finishing industries compliance only firms were shown to participate predominantly in waste management initiatives with little participation in market driven initiatives. In the fruit and vegetable processing industry there was a distinct difference between countries in the type of initiative that compliance only firms undertook. In the British/Irish sample this category of firm focused on market driven initiatives such as avoiding chemical ingredients, regional sourcing and organic raw materials. In contrast in the German sample, compliance only firms were engaged in mostly waste separation and packaging with limited participation in market initiatives. Italian compliance only firms chiefly undertook waste separation and energy efficiency initiatives. A common feature of the compliance only firms in all the countries was the lack of participation in environmental management initiatives. This was also a facet of both the furniture and textile finishing sectors.

Compliance plus firms in all the countries were engaged in a more extensive range of initiatives but to a much lesser extent than the excellence performing firms. In the UK/ROI the features that distinguished the compliance plus firms from the excellence firms were that compliance plus firms carried out fewer efficiency initiatives, had minimal participation in waste water treatment and limited take up of management initiatives. In Germany compliance plus firms lacked involvement in the market driven and management initiatives and to some extent initiatives associated with water. Italian compliance plus differed from the excellence firms in that proportionately fewer compliance plus firms undertook water efficiency initiatives and the market driven initiatives associated with artificial and organic materials.

In the FTF survey compliance only firms did not show the level of involvement in the market initiatives (particularly the usage of organic raw materials and regional sourcing) relative to the other initiatives as had been evident from the PQ data. There was a high degree of adoption of most initiatives irrespective of performance group and there was no significant difference in the take up of waste separation, packaging, energy efficiency, or environmental management initiatives between the performance groups. The initiatives that most clearly differentiated the compliance plus firms from the compliance only firms were environmental communication, water use and protection, and as referred to above, the market initiatives. In Germany as with the PQ results, the excellence performance firms were distinguishable from the other performance groups by the fact that they had a high degree of participation in the management initiatives, EMS and communications, and the market driven initiatives especially organic raw materials and regional sourcing.

In the UK/ROI there was no statistical difference in the size of the firm and the ability to undertake the market driven initiatives. Small firms were as likely as large firms to introduce initiatives to limit the use of artificial ingredients, and to increase the usage of organic raw materials and regional sourcing. As size increased proportionately more firms undertook efficiency and management initiatives. In Germany, as firm size increased, proportionately more firms were involved in management and market initiatives. Small firms in Germany concentrated on waste and packaging initiatives. In Italy there was no size effect associated with the type of initiatives undertaken.

The impetus to separate waste was a mixture of cost and regulation. The main motivation for introducing energy efficient initiatives was to achieve cost savings and firms that had invested in such initiatives did acknowledge a reduction in costs and hence increased profitability. In the UK/ROI and Germany the main reason for using environmentally friendly packaging was the demands of customers whilst in Italy this initiative was very much regulation driven. In all the countries, customer demands were the main motivator for avoiding artificial ingredients in the finished product. In addition, in Germany there was legislation prohibiting the use of colourings, flavourings and preservatives. The use of organic raw materials was similarly market driven. Environmental management and environmental communications were driven by the market and regulation.

In Germany and the UK/ROI the economic impact of initiatives was extremely positive in the FTF interviews but for Germany more marginal, from the evidence of the PQ data. In Italy the overall effect was only marginally positive. There was some variability in effects within and between countries for the same initiative.

When comparisons were made between environmental performance groups on the combined UK/ROI and Germany PQ database, there was evidence that improved environmental performance was associated with increments in economic performance. Small and large firms reported more positive economic impacts on average per initiative than middle-sized firms sampled. There was little net gain in labour in the UK/ROI, rather more in Germany and most in Italy.

Conclusions on findings on economic drivers and effects of individual initiatives

For all three sectors there was evidence of a hierarchy of initiatives performed by firms when comparisons were made between compliance only, compliance plus and excellence environmental performers. In furniture and textiles this implied a rise in the technical complexity of the adopted initiative and the driver (regulation for compliance only firms). In the case of fruit and vegetable processing, in the UK/ROI, there were initiatives that were adopted by compliance only firms, which were market driven, in Germany and Italy the type of initiative was less differentiated from those of compliance only firms in the other two industries. Excellence performers in all three sectors were distinguished by the extent they undertook management initiatives. The ability to take on these initiatives, as we have seen from the competitiveness considerations was not related systematically to R&D capacity or skills. There was some indication of a size effect on initiatives adopted, for example management initiatives were adopted above the very small firm size threshold. In the German fruit and vegetable processing sector, as size increased more firms were involved in market initiatives. In furniture in the UK/ROI as size rose more firms undertook solvent reduction. Nevertheless for the majority of the initiatives studied there was no size threshold.

Drivers for the adoption of initiatives were similar, for the same initiatives across the three industries, but there were differences across countries. In Italy regulation predominated, in UK/ROI regulation and cost were important, in Germany market pressures were much more important in bringing forth environmental initiatives

Analysis of the economic impact of initiatives indicated variability between positive and negative effects for the same initiative between firms. There was a great deal of variability in furniture but less in textiles and fruit and vegetable processing. The outcome was that the initiative produced a marginal economic benefit, on average, for furniture, and increasingly more positive for textiles and fruit and vegetable processing.

Comparing across the environmental performance groups, the furniture sector showed a marginal economic payoff reported by those firms which had moved above compliance only status, in the cases of textiles and fruit and vegetable processing the data clearly showed that economic benefits rise through the performance groups.

By size smaller firms reported fewer payoffs in furniture and textiles than larger firms. For small firms, the reverse was the case for the fruit and vegetable sector. In all three industries there were small positive effects on employment.

III. Obstacles to the adoption of environmental initiatives

A set of economic factors which may inhibit the adoption of environmental initiatives were considered viz. difficulties in raising capital, risk, regulatory uncertainty, pay back periods and so forth. These obstacles are those stated by respondents and not measured by the research.

Furniture

There was reasonable agreement between the countries that the most important constraints on the implementation of environmental initiatives into businesses were (i) the difficulties faced in raising capital for a new investment and (ii) that the regulations were too uncertain to allow managers the scope to plan for putting new technologies into practice. The latter was a reflection of a general shortfall in both knowledge about new technologies and the availability of information concerning current or proposed legislation. The British/Irish managers reinforced this perception by frequently citing the shortage of good quality advice. The opposite was the case for the Italian and German companies. Here the matter of receiving good quality advice was not considered to be a major obstacle although for the Italian firms specific, targeted advice from consultancy services was expensive.

Italian and British/Irish firm managers also considered that from a management point of view there were too many other priorities and hence environmental issues came low on a list of their concerns. In addition, the British/Irish firms believed that the shortage of management time to address such issues was an important constraint.

When these obstacles were broken down by environmental performance group, in Germany the top three obstacles were ranked in the same order irrespective of performance group. These were (i) problems in finding the necessary capital, (ii) the uncertainty of regulation and (iii) the unproven nature of clean technology. In the UK/ROI compliance only firms were mostly worried about acquiring the necessary capital, the imperfect nature of advice, demands on management time and business concerns. In contrast, compliance plus firms considered the regulation and information issues to be the most important.

There was no association between the size of the firm and the ranking of the obstacles, small and large firms alike endorsed the above findings.

Textile finishing

In terms of the constraints to implementing environmental initiatives there was little agreement across all the countries. The exception to this was the difficulty faced in raising capital for a new investment, which was ranked in first position by all the countries. In the UK/ROI and Germany the concerns about the payback time on clean technology were considered important but not viewed as a major constraint in Italy. Shortage of the correct skills and expertise was an issue in Italy and the UK/ROI but to a much lesser extent a concern in Germany. Italian and British/Irish firm managers also considered that from a management point of view there were too many other priorities and hence environmental issues came low down the list of concerns. Furthermore, the British/Irish firms believed that the shortage of management time to address such issues was a primary constraint. The latter was not rated as being a problem in either of the other two countries. Italian companies were concerned about the lack of good quality advice from both agencies and suppliers. The perception that regulation did not encourage firms to adopt clean technology was a concern in all the countries and the uncertainty of regulation in planning for technology changes was problematic in Germany and in the UK/ROI.

A breakdown of these obstacles by environmental performance group indicated for the UK/ROI, there was little to distinguish between the two performance groups. Both groups were concerned with finding capital for investment, management issues, and skill shortages although the compliance only firms were more anxious about the risky nature of regulation. The German responses by performance group showed little difference between the groups. Capital and payback were equally important, as was the uncertainty attached to environmental legislation, for planning for the future. Only the excellence group of firms were not worried about introducing clean technology despite its perceived risky nature. Italian firms did show some variability between the groups especially for payback, profit and the cost of environmental consultancy services.

There was no association between the size of the firm and the ranking of the obstacles, small and large firms alike reflected the general pattern.

Fruit and vegetable processing

In terms of the constraints to implementing environmental initiatives there was a general agreement across all the countries that the main limitation was the difficulty faced in raising capital for a new investment and the length of the associated payback period. Shortage of the correct skills and expertise was an issue in Germany but was of little or no concern in Italy or the UK/ROI. British/Irish firm managers considered that from a management point of view there were too many other priorities and hence environmental issues came low down the list of concerns. In addition, the British/Irish firms believed that the shortage of management time to address such issues was an important constraint. Management concerns were not considered to be serious obstacles in either Italy or Germany. The uncertainty of regulation in planning for technology changes was important in Italy and Germany as was the related issue concerning regulation and the lack of encouragement to firms to adopt clean technology. These obstacles that were allied to regulation were not important in the UK/ROI. Quality and cost of advice from external sources were not thought to be significant in any of the countries.

A breakdown of these obstacles by environmental performance group showed that with the exception of the profit motive, there was little to distinguish between the two performance groups in the UK/ROI. Both groups were concerned with finding capital for investment, payback and management issues. The compliance plus firms were more anxious about the need for the firm to make a profit irrespective of other concerns. With the exception of the problem associated with the cost of consultancy services, the German responses by performance group showed little difference between the groups. Compliance only firms were more concerned than firms in the other performance groups about the cost of obtaining advice. Italian firms showed little variability between the groups except for management priorities where compliance only firms considered this factor a severe restriction. There was no association between the size of the firm and the ranking of the obstacles, small and large firms alike endorsed the general pattern.

Conclusions on obstacles to the adoption of environmental initiatives

First, there was no relationship between size of firm and type of constraint for any of the sectors.

Second, what was most important in all sectors was a capital constraint. The opportunity cost of undertaking the initiatives, i.e. management priorities was a further common constraint, as was having the correct skills and expertise or advice. In furniture alone regulatory uncertainty and the unproven nature of clean technology was highlighted as a constraint. There was variation in the importance of these factors between countries.

Third, there was little to distinguish the constraints reported by any of the environmental performance groups. This may imply that these factors are always a constraint on environmental performance independent of the level of performance achieved.

IV. Environmental and Competitive Strategy

Analysis of the firm's environmental strategy (e.g. compliance with regulation, promotion of eco products, cost savings) was related to the firm's environmental performance. Do firm's pursuing an environmental market strategy, such as focusing on eco products, have strong environmental performance? The related question is whether firms following a cost leadership, product differentiation or niche product strategy have a different environmental performance. Does the competitive pressure implied by cost leadership tend to limit environmental performance to those initiatives driven by regulation and eco efficiency? These two analyses were brought together by asking the question is there an interrelationship between environmental and economic strategy?

Environmental strategy

Furniture

In terms of differences in company environmental strategies between the three performance groups, the Italian data showed no differences i.e. companies adopted a regulation strategy irrespective of their environmental performance. In the UK/ROI the only strategy to exhibit a significant difference between the performance groups was that of promoting eco-products with the excellence firms surpassing the compliance only and compliance plus firms. In Germany there was a significant difference for both regulation and eco-practice strategies. As firms moved beyond compliance the favoured strategy moved away from being regulation focused to the market centred strategy of improving brand image by undertaking environmentally friendly practices.

The responses from the FTF interviews regarding the strategy adopted by the company showed that in the case of Italy the conclusions were the same as for the PQ results i.e. an exceptionally high proportion of firms have endorsed regulation as their environmental strategy. In the UK/ROI and Germany the domination of regulation was still evident but not as markedly so. As with the findings from the PQ, there was clear evidence that when firms move beyond compliance they adopt environmental strategies that are proactive in terms of improving the firm's market position by differentiating themselves from their competitors through eco-practices or eco-products. These differences between compliance and beyond compliance firms were statistically significant.

Textile finishing

In all the countries the majority of firms that responded to the PQ viewed complying with regulations as the principal focus of the environmental strategy. There was however variation between the countries in the proportion of firms adopting this approach. Italian firms, irrespective of environmental performance group, showed a widespread adoption of a regulation strategy. German firms had the lowest proportion of firms endorsing a regulation strategy and a cost strategy was considered to be almost as important. This was not unexpected considering that many of the clean technology initiatives were adopted for cost reasons. In addition, in Germany there was reasonable support for the market driven, proactive environmental strategies of promoting eco-practices and eco-products. In the UK/ROI, complying with regulations was the prime strategy but it was combined with cost savings and the promotion of eco-practices to give a mixed approach. None of the countries showed any evidence of differences in the strategy pursued by either company size or environmental performance group.

Responses from the FTF interviews regarding the environmental strategy adopted were broadly the same as for the PQ results i.e. a high proportion of firms had endorsed regulation as their environmental strategy with little emphasis on environmentally proactive strategies. In the UK/ROI the company environmental strategy was almost equally divided between regulation and cost savings. Proactive strategies were virtually absent from the responses in contrast to the PQ, where there was a tendency towards eco-practices particularly for the excellence firms. The German results very much confirmed the pattern from the PQ with a mix of strategies. There was some limited evidence that when German firms move beyond compliance they adopted environmental strategies that were proactive in terms of improving the firm's market position by differentiating themselves from their competitors through eco-practices or eco-products. However, these differences between compliance and beyond compliance firms were not statistically significant.

Fruit and vegetable processing

On the basis of the PQ results there was variation between the countries in strategies that were followed. In the British/Irish sample firms, regulation was the preferred strategy of about half the companies surveyed, followed by eco-practice and eco-products. Cost strategies accounted for about 20% of responses. In an industry where the environmental initiatives were driven by a combination of regulation, market and cost this was to be expected. German companies adopted regulation, cost and eco-practice strategies in almost equal proportions. As in the UK/ROI this was a reflection of the initiative drivers. In Italy firms adopted either a regulation or eco-practice strategy and it was unexpected that cost as a strategy did not feature more notably considering that almost 40% of the firms adopted cost driven energy saving initiatives.

In terms of difference between the environmental performance groups, in Germany there was a significant difference across groups in the adoption of strategies to promote brands by eco-practice or develop new markets for eco-products with the excellence firms preferring these strategies. In Italy, compliance only firms favoured regulation whilst compliance plus and excellence firms showed a distinct preference for adopting good environmental practice as the primary strategy. In the UK/ROI there was no significant difference in the strategy adopted across performance groups.

From the FTF interviews in all the countries the results differed to a certain extent from the findings of the PQ. Cost saving as an environmental strategy became more important in the UK/ROI, especially amongst the compliance only firms, and Italy whilst the influence of regulation diminished. Except in Italy, the emphasis on environmentally proactive strategies, particularly eco-products was somewhat diminished. The German results showed that compliance only firms did not follow a mix of strategies as shown in the PQ but were only involved in regulation or cost savings strategies. There was evidence that when German firms move beyond compliance they adopted environmental strategies that were proactive in terms of improving the firm's market position by differentiating themselves from their competitors through eco-practices or eco-products. However, these differences between compliance and beyond compliance firms were not statistically significant.

Conclusions on environmental strategy

Regulation was an important strategy irrespective of the environmental performance of the firm. German firms showed evidence in furniture and fruit and vegetables for their environmental strategies becoming more proactive as they moved along the environmental performance groups, to incorporate eco-practices, eco-products and markets. There was some indication of a similar finding for the UK/ROI and Italy respectively, in furniture and fruit and vegetable processing. Other than this, although there was a representation of different environmental strategies, in different proportions, in each sector, there was little evidence of an association between environmental performance and environmental strategy. The latter may reflect the mix of environmental drivers that all firms face.

Competitive strategy

Furniture

At the FTF interviews the managers were asked to describe the company's competitive strategy. On the basis of the responses the firms were characterised as belonging to one of the following: cost leadership with an emphasis on low cost, differentiation where the emphasis was on uniqueness, or niche with a focus on a particular segment of the market. In the UK/ROI the firms interviewed adopted all three strategies in almost equal proportions. There was a difference between the two environmental performance groups in the type of strategy adopted. Cost leadership was the approach adopted by almost half of the compliance only firms compared with about one-quarter of the compliance plus firms. For firms following a differentiation strategy these proportions were reversed, and reduced for compliance only firms. In Italy the picture was similar to the distribution among the British/Irish compliance only firms. Since all the Italian firms except one were categorised as compliance only this position was not unexpected. In Germany the situation was somewhat different with no firms pursuing a low cost competitive strategy and a high proportion implementing a niche strategy irrespective of environmental performance.

Compliance plus firms were more likely to be pursuing a non-cost based competitive strategy. This may be because, their ability to charge a price premium allows them the flexibility to adopt environmental initiatives or it may be that they consider that undertaking environmental initiatives reflects a form of differentiation/quality. Thus there was an expectation that firms following a cost leadership strategy would undertake fewer initiatives than those with a differentiation or niche strategy. However, the data did not support this assumption.

Textile finishing

In the UK/ROI of the firms interviewed, almost half adopted a differentiation strategy with roughly equal proportions following either a cost or niche competitive strategy. In Italy the picture was fairly similar to the distribution among the British/Irish firms. In Germany the situation was somewhat different with only one firm pursuing a low cost competitive strategy and equal proportions adopting differentiation or niche strategies. *A priori*, it was anticipated that the better environmental performing firms would be clustered in the niche category but this was not the case for any of the countries. In fact, there was no difference between environmental performance groups and the type of strategy adopted. Also the expectation that the firms that pursued a cost leadership would undertake fewer initiatives than those with a differentiation or niche strategy was not proven.

Fruit and vegetable processing

In the UK/ROI almost half of the firms followed a cost leadership strategy with about one-third adopting a differentiation strategy and the remainder a niche strategy. In Italy about half of the firms had a differentiation strategy, one-third niche and the remaining one-sixth a cost strategy. In Germany the situation was again different with niche strategies dominating followed by differentiation and less than 12% of firms pursuing a low cost competitive strategy. *A priori*, it was anticipated that the better environmental performing firms would be clustered in the differentiation or niche categories but this was not the case for the UK/ROI where proportionately more compliance plus firms adopted a low cost strategy compared with compliance only firms. The differences between these two performance groups was statistically significant. In Italy and Germany there was no difference between the environmental performance groups regarding the type of strategy adopted. Also the expectation that the firms that pursued a cost leadership strategy would undertake fewer initiatives than those with a differentiation or niche strategy was evident only in Italy.

Environmental and competitive strategy

Furniture

When comparing the relationship between the competitive strategy and the environmental strategy adopted by the firms there was an assumption that firms with a differentiation or niche competitive strategy would have had an environmental strategy that reflected this approach i.e. they would have implemented environmental strategies relating to environmentally friendly practices and products. This was not the case and environmental strategies driven by regulation and cost savings were the predominant approaches regardless of the company's competitive strategy.

Textile finishing

The data showed that in the UK/ROI and Italy environmental strategies dominated by regulation and cost savings were the predominant approaches regardless of the company's competitive strategy. In Germany where there was only one firm with a cost leadership strategy, there was some limited evidence that companies with a differentiation or niche strategy did also embrace eco-strategies as well as those focused on cost savings and regulation.

Fruit and vegetable processing

There was evidence in Italy and Germany that firms with a differentiation strategy did implement proactive environmental strategies. However, this was not so with firms that had focused competitive strategies where regulation and efficiency strategies were undertaken by at least half of these firms. There was no evidence of a relationship for the UK/ROI.

Conclusions on environmental and competitive strategy

There was some limited evidence to suggest a link between environmental and competitive strategy, namely that product differentiation and the pursuit of niche markets facilitated environmental performance. As expected there were many reasons why a firm will adopt alternative competitive strategies independent of their environmental performance, but there was slight evidence that strong environmental performance can be associated with differentiated and niche strategies. The important conclusion that follows from this is that cost leadership strategies are not a constraint on environmental performance.

9.2.3 Hypothesis 2

The role of management and the culture of the business organisation is important to the take up of environmental initiatives.

The research hypothesised that the internal 'culture' of the firm was important in effecting environmental decision-making and environmental performance. Attitudes were measured using a questionnaire to manufacturers during the face-to-face interviews (some questions were also asked in the postal questionnaire). This enabled an analysis to be made primarily based on differences in attitudes compared with firm environmental performance.

What can be said? First, it was clear that attitudes towards the environment were on average positive. Irrespective of the firm's own environmental performance, managers were positive about the benefits of protecting the environment.

Second, analyses of the responses revealed that environmental attitudes were not related to the environmental performance of sample firms in general (there was some significant correlation between environmental performance and environmental attitudes in the case of the Italian fruit and vegetable processing sector).

Third, there was no discernible relationship between a firm's pro environmental stance or cultural score and a business strategy focused on environmental performance (e.g. eco-products or eco-efficiency) either within sample countries or across sample countries.

Fourth, most respondents did not believe that environmental policy had a deleterious effect on competitive performance, although German firms were more likely to note a negative effect.

Fifth, there was no clear effect of size on the environmental attitudes of the firm.

9.2.4 Hypothesis 3

The dissemination of information about environmental initiatives is likely to be a critical determinant of diffusion

A third key area of interest was the importance and influence of information and advisory sources, in enabling the take up of clean technologies.

SME suppliers, customers, universities, research associations, professional associations, public bodies and private consultancies were considered. The investigation examined the advisory process (between firm and advisor); the drivers and obstacles to seeking advice; the impact of the advisory services; and from the standpoint of the manufacturing firm, those advisory services they considered most useful.

Findings

There was an abundance of easily accessible and cheap information, indeed no manufacturing firm complained of a lack of advice as an obstacle to the adoption of clean technology or environmental initiatives (except in the case of UK/ROI's furniture sector).

However, most SMEs remained unaware of the environmental impacts associated with their business activities and had a low awareness of environmental programs or resources available to them. As environment was not regarded as important, many SMEs did not feel the need for any advisory services. Moreover SMEs were isolated, they did not have the networks to understand and evaluate technology. Drivers were important especially environmental regulation.

The most common support services included: free or subsidised briefings, seminars, workshops targeting specific business activities or environmental requirements and guiding SMEs through tailored, and affordable environmental programs and staff training. The sources of advice admitted that support needs varied enormously thus every enquiry was dealt with on an individual basis.

There was a mix of methods used to obtain clients including referrals, advertising etc. Low responses and poor interest in free advisory services were notable in the UK/ROI. This was compounded because the quality of advice was stated to be problematic in both the public and private sectors. In Germany firms felt there was not much difference in the quality of service offered by public and private sectors, and in Italy, firms observed that public sector services were more reliable and efficient than those of the private sector.

The SOAs stated that the demand for their services usually reflected five key drivers for SMEs:

- Specific environmental legislation or regulatory matters.
- Supply chain pressure.
- Customers demanding proof of environmental management performance.
- Financiers or insurers seeking assurances regarding environmental risk.
- Specific environmental concerns or problems.

The solutions most frequently recommended by consultants to their clients were implementing changes in the production process, introducing technical monitoring of production, providing technical answers to product mismanagement or assistance in innovation processes and improving product quality. Almost all the respondents maintained that, in order to improve their environmental performance, small firms needed to reorganise production processes and adopt new machinery, compatible with higher environmental standards.

Most interviewees perceived that their advice substantially contributed towards some change in the firm's environmental position. This may be a result of implementing an Environmental Management System or other input orientated system, which creates information and awareness of adopting various initiatives. Some of the benefits realised by firms as a result of introducing EMS were cited as: greater awareness of environmental responsibilities among the workforce, better relationships with regulatory bodies, insurers and the local community and better internal management and communication. It was also pointed out, that the person responsible for environmental initiatives was more important than the initiative itself for the realisation of a substantial shift. A positive shift that impacted on the firm's environmental culture was universally reported.

Increased economic efficiency, as an important measure of the advisory activity of SMEs was also investigated. Though most SOAs involved in production and technical changes in SMEs were unable to give figures, in the UK/ROI and in

Germany almost everybody interviewed and half of the those at Italian advisory firms claimed that their involvement led to a significant increase in firms' economic efficiency.

Obstacles to the take up of advice by firms (as identified by SOAs)

The following were highlighted as important factors (while acknowledging that there was a wide variability across SMEs).

Resources

Smaller companies suffered from a shortage of resources. The people resource constraint was more important than any financial constraints (though lack of capital could also be important). The smaller the firm the more tasks fall to one particular person. For example the H&S manager, the production manager or quality manager may get the job of addressing environmental issues. There was no single person who can specialise in the environment alone. Time to consider and time to implement were very important. Firms may not explicitly calculate the costs of their time, but they were very aware of time constraints.

Resistance to change

Many firms were not open to change and management may even have an obstructive agenda. Often H&S managers were insufficiently powerful to be able to push though the responsibility they have been given. Smaller firms needed more counselling than they asked for and smaller firms were not aware of what initiatives could be carried out and how to do this.

Cost focus

There was scepticism about the extent to which opportunities for cost savings exist, and therefore small firms were rarely persuaded by cost-saving opportunities. Because of the cost there was a resistance to the idea of using consultants.

Culture

The resource constraint was combined with the culture within the firm, namely the view that environmental concerns were a necessary evil, and dealt with through compliance. The result was that environmental responsibility was often combined with health and safety and not viewed as a separate issue. This reinforced the low priority, and often led to environmental improvements being treated as 'interruptible processes', that take second place to production related activities.

Firms did not look for the most efficient way of managing environmental concerns, but instead they focused on compliance. This suggests that it requires a change of mindsets to see the environment as another business issue. Firms did not view environmental performance as being directly related to competitiveness;

therefore the corporate culture has an important role in supporting and sustaining the firm's commitment to continuous environmental improvement.

Management attitude was conceived to play an emphatic role in the environmental performance of firms across all the countries. This was contrary to the findings (except for the Italian fruit and vegetable processing sector) of the measured relationship between environmental attitudes and environmental performance.

Best sources of advice (as perceived by the SMEs)

Small and medium-sized enterprises favoured a wide range of sources as information for clean technologies and environmental initiatives for companies like themselves. There was wide variation by sector and country. Noted examples were professional associations, fellow companies, government bodies and regulators, suppliers of machinery and industry associations, customers and consultants. Hence all sources of advice were relevant to the search for clean technologies.

To summarize, sources of information and advice were important. The diversity was important and the main difficulty was with their under use.

9.3 Implications of the Findings for Policy

The study has focused in depth on three hypotheses expected to influence SME environmental performance, competitiveness, culture, and the availability of advice.

The hypothesis with respect to competitiveness acknowledged two-way causation. Is it the case that the more competitive SMEs have a greater capacity to adopt environmental initiatives? The answer appears to be no. Despite competitive advantages which may variously take the form of above average profitability, growth and associated R&D, skills and modernity, there was only scattered evidence to suggest any of these was important. The policy implication follows that strong environmental performance is not constrained by the initial competitive status of the firm. This finding was further underlined by the evidence shown that firm constraints on the adoption of environmental initiatives did not differ according to environmental performance groups. Capital or skills were a constraint irrespective of the level of environmental performance already achieved. However there was evidence that environmental performance can be a function of the size of the firm, and that was demonstrated for a number of environmental initiatives, notably environmental management, but also some technical initiatives.

There was also evidence that improvements in environmental performance were associated with increasing technical complexity followed by the adoption of environmental management initiatives.

There was no evidence that a firm's competitive strategy constrained its environmental strategy. The pursuit of cost efficiency did not militate systematically against the environmental goals and performance of the firm. Similarly there was evidence to indicate that firms could use environmental strategy to differentiate themselves, to create a competitive advantage in the market.

At the aggregate level, the finding that firm competitiveness was generally unrelated to environmental performance implied also that strong environmental performance was not associated with a weakening of competitive performance.

9.3.1 Competitiveness and Economic Effects

This was tested by examining the economic effects of individual initiatives. In general the majority of economic impacts were positive but more so for firms in the textiles and fruit and vegetable industries. There was some evidence that for two sectors payoffs were smaller for very small firms. Moreover the ratio of positive economic benefits arising from any initiative rose as the firm moved up the environmental performance ladder. This supports the finding that improved environmental performance is not associated with a worsening of economic performance.

However this analysis had an important policy implication. It indicated that both across countries, across sectors and within sectors, there was a variety of positive and negative economic experiences from adopting environmental initiatives. On average the experience was positive but for any single firm there was a probability that the experience could be negative compared with their current techniques and performance. It does not therefore follow that all firms can improve their environmental performance with, uniformly, positive economic outcomes.

This variability in experience is likely to be an important barrier to movement up the environmental improvement ladder in response to any given driver, with the exception of regulation.

9.3.2 Obstacles

The obstacles noted by the firms themselves were primarily a capital constraint, payback, priority and management time and an absence of the correct skills and/or advice. It was notable that these obstacles were neither a function of the size of the firm or the environmental performance of the firm.

The obstacles stated by firm advisors were broadly similar. A lack of resources (both time and capital), scepticism about the benefits to be derived from the environmental initiatives, and rather differently, resistance to change, and the view that environmental performance was a necessary constraint on the activities of the firm and an interruption to production related activities.

The advisors also highlighted the lack of demand for their services (as indicated by the low take up of even free services). They emphasised the need for external expertise, especially where the technicalities of the environmental initiative was outside the scope of the business of the SME. Like the SMEs themselves they would emphasise the need for correct skills but a lack of take up of those skills.

This may be a consequence of the SMEs preferred choice of advisor. These were various and included industry associations, fellow companies, regulators, customers, consultants and suppliers, each with different types of expertise.

Hence the hypothesis that information about environmental initiatives is likely to be a critical determinant of diffusion is supported particularly in the context of the provision of relevant skills and firm specific advice.

9.3.3 Environmental Performance and Environmental Culture

A widely quoted barrier to the take up of clean technologies is the firm's culture. This was emphasised in many of the interviews with advisors and is one of the three important hypotheses considered in this study.

Investigation of the hypothesis indicated that positive environmental attitudes were quite common, and were positively held irrespective of the firm's own environmental performance, or environmental strategy. This positive view was independent of the size of the firm. Moreover most respondents (with the exception of many German counterparts with already good environmental performance) did not believe that environmental improvement had a deleterious effect on the economic performance of the firm.

9.3.4 Research and Policy Conclusions

What does all this boil down to? Policy should not be constrained by the expected impact on competitiveness of pressures to improve the environmental performance of firms. However it must be recognised that similar pressures can lead to different economic impacts on what look like identical firms. Firms need more expert help than they seek or realise they need to adopt environmental initiatives. Poor environmental performance is not the outcome of a negative attitude to the environment.

References

Advisory Council on Science and Technology (ACOST) (1992) Cleaner technology. HMSO, London

Antle R, Eppen G (1985) Capital rationing and organizational slack in capital budgeting. Management Science 31: 163–174

Arrow K (1962) The economic implications of learning by doing. Review of Economic Studies 29: 155–173

Ashford NA (1993) Understanding technological responses of industrial firms to environmental problems: Implications for government policy. In: Fischer K and Schot J (eds) Environmental strategies for industry. Island Press, Washington DC

Bailly AS (1987) Service activities and regional development: Some European examples. Environment and Planning A 19: 653–668

BCC (1994) Small firms survey: Environment the view of the small firm. British Chamber of Commerce/Alex Lawrie, London

Blackman A, Bannister GJ (1996) Community pressure and clean technologies in the informal sector: An econometric analysis of the adoption of propane by traditional brickmakers. Journal of Environmental Economics and Management 35: 1–21

CBI (1994) Environment costs. Confederation of British Industry, London

Christie I, Northcott J, Walling A (1990) Employment effects of new technology in manufacturing. PSI, London

Christie I, Rolfe H, Legard R (1995) Cleaner production in industry. Integrating business goals and environmental management. PSI, London

CIA (1993) Survey of investment intentions. Chemical Industries Association, London

Coffe WJ, Polese M (1989) Producer services and regional development: A policy orientated perspective. Papers of the Regional Science Association 67: 13–27

Commission of the European Communities (1983) Cleaner technologies in the EEC cement industry. CCE EUR 8394 EN. Office for Official Publications of the European Communities, Luxembourg

Commission of the European Communities (1984) Cleaner technologies in the chemical industry. ERL CCE EUR 8442 EN. Office for Official Publications of the European Communities, Luxembourg

Commission of the European Communities DG XI (1993) Towards sustainability. Office for Official Publications of the European Communities, Luxembourg

Commoner B (1994) Pollution prevention: Putting comparative risk assessment in its place. In: Finkel A and Golding D (eds) Worst things first? The debate over risk-based national environmental priorities. Resources for the Future Communication Association, Washington, DC

Cramer J (1998) Environmental management: From 'Fit' to 'Stretch'. Business Strategy and Environment 7: 162–172

Cropper ML, Oates WE (1992) Environmental economics: A survey. Journal of Economic Literature 30: 675–740

De Marchi B (1995) Information to the citizens in the European Communities legislation concerning environmental policy, nuclear energy and radiation protection. Technical Note no I.95.142. European Commission Joint Research Centre, Ispra, Italy

Department of the Environment (1994) Top management's attitudes to energy and the environment. HMSO, London

Dixit A, Pindyck R (1994) Investment under uncertainty. Princeton University Press, Princeton, NJ

ECMT/OECD (1994) Internalising the social costs of transport. ECMT/Organisation for Economic Co-operation and Development, Paris

Egri CP, Pitfield LT (1996) Organizations and the Biosphere. In Clegg S, Hardy C, WR Nord (eds) Ecologies and Environments. Handbook of Organization Studies, Sage, London

Ekins P, Speck S (1998) The impacts of environmental policy on competitiveness: Theory and evidence. In: Barker T, Köhler K (eds) International competitiveness and environmental policies. Edward Elgar, Cheltenham, pp 33–70

ENDS (1994) Integrated Pollution Control: The first three years. Environmental Data Services Ltd, London

Ester P (1991) Environmental concern in the Netherlands. In O'Riordan T and Turner RK Progress in research management and environmental planning, Vol. 3. John Wiley and Sons, London

Etzioni A (1988) The moral dimension. Free Press, New York

Fairbairn B (1997) ISO9000 and organizational culture. Paper presented at the Institute of Quality Assurance, 4 February 1997, Manchester

Fineman S (1997) Constructing the green manager. British Journal of Management 8: 31–38

Gerrans PA, Hutchinson WE (1998) EMS and SMEs: Current and likely impact of environmental management systems (EMS) on small to medium enterprises (SMEs). Paper presented to the 43rd ICSB World conference on Entrepreneurship, Nanyang Technological University, Singapore, 8–10 June

Gillis WR (1987) Can service producing industries provide a catalyst for regional economic growth? Economic Development Quarterly 1: 249–256

Glass N (1996) Competitiveness and the Environment. Paper presented to conference on Environmental Economic Policies: Competitiveness and Employment. October, Dublin

Granovetter M (1985) Economic action and social structure: The problem of embeddedness. American Journal of Sociology 91

Green K, McMeekin A, Irwin A (1994) Technological trajectories and R&D for environmental innovation in UK firms. Futures 26: 1047–1059

Greenberg R, Unger C (1992) Performance indicators of environmental processes Total. Quality Environmental Management, Summer: 405–408

Hartje VJ, Lurie RL (1985) Research and development incentives for pollution control technologies. International Institute for Environment and Society, IUUG, Berlin

Heath A, Jowell R, Curtice J, Evans G, Field J, Witherspoon S (1991) Understanding political change: The British voter 1964–1987. Pergamon Press, Oxford

Hitchens DMWN, Farrell F, Lindblon J, Triebswetter U (2001) The Impact of Best Available Techniques (BAT) on the competitiveness of European industry. European Commission Joint Research Centre, Report EUR 20133EN, Seville, Spain

Hodge I (1995) Environmental economics. Macmillan, London

Hutchinson A, Chaston I (1994) Environmental management in Devon and Cornwall's small and medium sized enterprise sector. Business Strategy and the Environment 3: 15–22

Illeris S (1989) Services and regions in Europe. Gower, Brookfield, VT

Irwin A, Hooper P (1992) Clean technology, successful innovation and the greening of industry. Business Strategy and the Environment 1: 13–24

Jacobson D, Andréosso-O'Callaghan B (1996) Industrial economics and organisation. McGraw Hill, Maidenhead

Jaffe AB, Peterson SR, Portney PR, Stavins RN (1995) Environmental regulation and the competitiveness of US manufacturing: What does the evidence tell us? Journal of Economic Literature 33: 132–163

James P (1994) Business environmental performance measurement. Business Strategy and the Environment 3:59–67

Kodama F(1991) Analysing Japan's high technologies: The technological paradigm shift. Pinter Press, London

Likert R (1932) A technique for the measurement of attitudes. Archives of Psychology 140 June

Likert R (1961) New patterns of management. McGraw Hill, New York

Ludevid AM (2000) Small and medium sized enterprises, perceptions of the environment: A study from Spain. In Hillary R (ed) Small and medium sized enterprises and the environment. Greenleaf, Sheffield, pp 61–74

Marshall JN (1988) Services and uneven development. Oxford University Press, Oxford

Marshall JN, Wood PA (1992) The role of services in urban and regional development: recent debates and new directions. Environment and Planning A 24: 125–70

Martinelli P (1991) Producer services location and regional development. In: Mouleart F and Daniels PW (eds) The changing geography of advanced producer services. Belhaven Press, London, pp 70–90

Medhurst MJ (1994) Macroeconomic aspects of environmental policies and competitiveness. In: Environmental policies and industrial competitiveness, Organisation of Economic Co-operation and Development, Paris

Meller E (1986) Industry experience in the implementation of clean technologies in the Federal Republic of Germany. Federation of German Industries/BDI, Köln

Merritt JQ (1998) EM into SME won't go? Attitudes, awareness and practices in the London Borough of Croydon. Business Strategy and Environment 7: 90–100

Newman B, Irwin H, Lowe K, Mostwill A, Smith S, Jones J (1992) Southern Appalachians wildland proposal. Wild Earth, Special issue 1:pp 46–60

O'Connor M (1997) The internalisation of environmental costs: Implementing the Polluter Pays Principle in the European Union. International Journal of Environment and Pollution 7: 450–482

OECD (1985) Environment policy and technological change. Organisation for Economic Co-operation and Development, Paris

OECD (1987) The promotion and diffusion of clean technologies. Organisation for Economic Co-operation and Development, Paris

OECD (1990) Promoting the development and dissemination of environmentally-friendly technologies. Organisation for Economic Co-operation and Development, Paris

OECD (1991) Policy tools to encourage environmentally friendly technologies in the 1990s. Report prepared for the OECD Technology and Environment Programme, presented to its Workshop on Environmentally Friendly Technologies, June 1991

OECD (1992a) Implementing environmental protection: Enterprise level issues. Organisation for Economic Co-operation and Development, Paris

OECD (1992b) Government policy options to encourage cleaner production and products in the 1990s. Organisation for Economic Co-operation and Development, Paris

OECD (1995) Technologies for cleaner production and products. Organisation for Economic Co-operation and Development, Paris

OECD (1999) Regulatory reform for smaller firms. Working paper, Directorate of Science, Technology and Industry. Organisation for Economic Co-operation and Development, Paris

Oosterhuis FF, Rubik F, Scholl G (1996) Product policy in Europe. Kluwer Academic Publishers, London

O'Riordan T (1985) What does sustainability really mean? Theory and development of concepts of sustainability. Centre for Economic and Environmental Development, London

PA Consulting Group (1991) Cleaner technology in the UK. HMSO, London

Pasquero J (1991) Supraorganisational collaboration: the Canadian environmental experiment. Journal of Applied Behavioural Science 27: 38–64

Perry M (1991) The capacity of producer services to generate regional growth: some evidence from a peripheral metropolitan economy. Environment and Planning 23:1331–1347

Petts J, Herd A, O'Eocha M (1998) Environmental responsiveness, individuals and organizational learning: SME experience. Journal of Environmental Planning and Management 41: 711–731.

Petts J, Herd A, Gerrad S, Home C. (1999) The climate and culture of environmental compliance within SMEs. Business Strategy and Environment 8: 14–30.

Porter M (1990) The competitive advantage of nations. Free Press, New York

Porter M, Van ,der Linde C (1995) Towards a new conception of the environment–competitiveness relationship. Journal of Economic Perspectives 9: 97–118

Roome N (1994) Business strategy, R&D management and environmental imperatives. R&D Management 24: 65–82

Schaper M (2002) Small firms and environmental management: Predictors of green purchasing in Western Australian pharmacies. International Small Business Journal 20: 235–251

Schmidheiny S (1992) Changing course: A global business perspective on development and the environment. MIT Press, Cambridge, MA

Simon HA (1962) New development in the theory of the firm. American Economic Review Papers and Proceedings, 52: 1–15

Smith MR (1977) The challenge of change. Cornell University Press, Ithaca, NY

Staudenmaier J (1985) Technology's story tellers. MIT Press, Cambridge, MA

Stead E, Stead J (1992) Management for a small planet. Sage, London

Stoneman P (1983) The economic analysis of technology change. Oxford University Press, Oxford

Storey DJ, Keasey K, Wynarcyzk P (1987) The performance of small firms: Profits, jobs and failures. Croom Helm, London

Tilley FJ (1998) The gap between the environmental attitudes and the environmental behaviour of small firms: With an investigation of mechanical engineering and business services in Leeds. PhD thesis, Leeds Metropolitan University, UK

Tilley FJ (1999) The gap between the environmental attitudes and the environmental behaviour of small firms. Business Strategy and Environment 8: 238–448.

Wallace D (1995) Environmental policy and industrial innovation. Earthscan/Royal Institute of International Affairs, London

Welford R (1994) Breaking the link between quality and the environment. Business Strategy and the Environment 2: 4

Welford R (1995) Environmental strategy and sustainable development: The corporate challenge for the 21st Century. Routledge, London

Whitaker B (1993) The business case for the environmental and the roles of legislation and economic instruments in enhancing the environmental performance of British business.

Environmental Management Working Group Papers, Advisory Committee on Business and the Environment/ ACBE, London.
White A, Becker M (1991) Total cost assessment: Catalyzing corporate self interest in pollution prevention. Paper presented to the National Academy of Engineering Workshop: Engineering our way out of the dump. Woods Hole, MA, July 1–3
White A, Becker M, Goldstein J (1992) Total cost assessment: Accelerating industrial pollution prevention through innovative project financial analysis. Office of Pollution Prevention and Toxics, EPA/741/R-92/002, May
White A, Savage D, Brady J, Cavander D, Lach L (1995) Environmental cost accounting for capital budgeting: A benchmark survey of management accountants. Office of Pollution Prevention and Toxics, EPA/742/R-95/005, September
Williamson OE (1963) Managerial discretion and business behaviour. American Economic Review 53: 1032–1057
Winter L, Ledgerwood G (1994) Motivation and compliance in environmental performance for small and medium sized companies: A model based on empirical evidence from a pilot investigation of small businesses in the English West Midlands. Greener Management International 7: 62–72
Witherspoon S, Martin J (1992) What do we mean by green? In: R Jowell et al. (eds) British Social Attitudes: The 9th Report. Aldershot, Darmouth, pp 1–26
Yakowitz H (1991) Issues effecting the usefulness of the environmental auditing processes a driving force for cleaner production. In Backman M and Lindqvist T (eds) Environmental auditing production strategies. Proceedings of an international expert seminar, Trolleholm Castle, Sweden

Index

MIX
Papier aus verantwortungsvollen Quellen
Paper from responsible sources
FSC® C105338

If you have any concerns about our products,
you can contact us on
ProductSafety@springernature.com

In case Publisher is established outside the EU,
the EU authorized representative is:
Springer Nature Customer Service Center GmbH
Europaplatz 3, 69115 Heidelberg, Germany

Printed by Libri Plureos GmbH
in Hamburg, Germany